Robert B. Gardner
University of North Carolina

The Method of Equivalence and Its Applications

SOCIETY FOR INDUSTRIAL AND APPLIED MATHEMATICS
PHILADELPHIA, PENNSYLVANIA 1989

Printed by Capital City Press, Montpelier, Vermont.

Library of Congress Cataloging-in-Publication Data

Gardner, Robert B.
The method of equivalence and its applications.

(CBMS-NSF regional conference series in applied mathematics ; 58)
Includes bibliographical references.
1. Geometry, Differential. 2. Invariants.
3. Transformations (Mathematics) I. Title. II. Title: Equivalence and its applications.
QA649.G36 1989 516.3′6 89-21972
ISBN 0-89871-240-8

Contents

v INTRODUCTION

1 LECTURE 1. Equivalence Problems

11 LECTURE 2. Lifting of Equivalence Problems to G-Spaces

19 LECTURE 3. The Structure Equations

31 LECTURE 4. Reduction of the Structure Group

47 LECTURE 5. The Inductive Step

57 LECTURE 6. e-Structures

71 LECTURE 7. Global Results and Involutive Structure

81 LECTURE 8. Serendipity

89 LECTURE 9. Normal Forms and Generalized Geometries

97 LECTURE 10. Prolongation

109 APPENDIX. Translation of the part of Cartan's paper "Les sous-groupes continus de transformations" on the equivalence problem

123 REFERENCES

Introduction

Élie Cartan's method of equivalence was a natural outgrowth of the program outlined by Felix Klein in 1872 in his famous lecture to the faculty at Erlangen. In this lecture entitled "Vergleichende Betrachtungen über neuere geometrische Forschungen" (A Comparative Review of Recent Research in Geometry), (see [Kl. 1921]) Klein developed the thesis that geometry was the study of invariants of group actions on geometric objects. It seems incredible to me that two nineteenth-century public lectures like Klein's, and Riemann's "Über die Hypothesen, welche der Geometrie zu Grunde liegen" (On the hypothesis upon which geometry is founded) (see [R. 1953]), could have such an effect on mathematics. Perhaps there is a lesson in history here that needs to be appreciated by modern mathematicians.

Lie, who was a good friend of Klein, rose to the challenge of finding a general method to uncover such invariants. Unfortunately, Lie's approach had some serious defects. The first problem arose because the determination of invariants always involved solving a system of first-order partial differential equations. Such an approach worked if special tricks could be applied, and many low-dimensional problems were solved using Lie's method. However, it was clear that solving systems of first-order partial differential equations was not an efficient way to solve problems with higher complexity. The second problem was that efficient determination of the infinitesimal generators, which defined the system of partial differential equations, normally required an explicit parameterization of the group involved. This may not seem like a problem, but an explicit parameterization of the orthogonal group in even five variables is not an easy task. A more dramatic example is an exceptional simple group like G_2, where standard algebraic tricks are not applicable.

Élie Cartan's approach was motivated by his work on infinite pseudogroups, which also fit into the Erlangen program, since Klein also envisioned invariants of structures under collections of diffeomorphisms like the set of all volume preserving or contact transformations. In 1908, in what has to be one of the most remarkable papers in mathematics, "Les sous-groupes des groupes continus de transformation," Élie Cartan formulated and described a procedure to find the invariants of geometric objects under pseudogroups defined by first-order partial differential equations. He did this in just 25 pages out of a 137-page paper.[1] The impact of part of these twenty-five pages on Klein's program was eloquently described by J. Dieudonne:

> Finally, it is fitting to mention the most unexpected extension of Klein's ideas in differential geometry. He had envisaged groups of isometries of Riemannian spaces as a possible field of study of his program, but in general a Riemannian space does not admit any isometries except for the identity transformation. By an extremely original generalization, É. Cartan was able to show that here as well the idea of "operation" still plays a fundamental role; but it is necessary to replace the group with a more complex object, called the "principal fiber space"; one can roughly represent it as a family of isomorphic groups, parameterized by the different points under consideration; the action of each of these groups attaches objects of an "infinitesimal nature" (tangent vectors, tensors, differential forms) at the same point; and it is in "pulling up to the principal fiber" that É. Cartan was able to inaugurate a new era in the study (local and global) of Riemannian spaces and their generalizations.[2]

The procedure was used with great success by Cartan and his students. In particular the names of Hachtroudi, Debever, Vranceanu, Švec, and Chern figure prominently as practitioners of the method in the 1930s, 1940s, and 1950s. Then there was a hiatus of twelve years until the 1960s when Singer, Sternberg, Guillemin, Kuranishi, Kodaira, and Spencer led a program to make Cartan's method rigorous and to answer basic questions about the applications of the method. Although many papers were written during these years and many basic results were established, such as the Kuranishi prolongation theorem, the development of the theory of prolongation of Lie algebras, and foundations of the theory of pseudogroups, no new significant geometric equivalence problems were solved. The reason, I believe, was that the method left too much apparent freedom in the way part of the constructions were done. In particular, the process of Lie algebra compatible absorption of torsion and the process of reduction of structure group were not laid out in any systematic way. In fact, reading most of the examples worked out by Cartan made you feel that special tricks and brilliant observations were part of the method. After thinking about this method for another twenty years and talking periodically with S.S. Chern and my students, especially Robert Bryant, I realized that mixing Cartan's original method with the concept of principal components in Cartan's theory of Répère Mobile led to an algorithmic way to execute Cartan's method. Although it is not explicitly stated, it is clear to me Cartan also must have been aware of this technique. This was first described in lectures I gave in Houghton, Michigan, in 1982 (see [G. 1983]) and has resulted in a flurry of new geometric equivalence problems being solved. In particular I mention the works of Bryant, Shadwick, Kamran, Olver, Wilkens, Grissom, D. Thompson, G. Thompson, and myself.

One may wonder why anyone would need twenty years to understand a paper. However, I was not alone in experiencing difficulty with parts of the theory of Répère Mobile and the method of equivalence, as the following two citations testify.

---

[1]A translation of this section of his transhistoric work is provided in the appendix.

[2]From Jean Dieudonné's introduction to *The Erlangen Program* by Felix Klein, translation by my Research Experience for Undergraduates assistant Adam Falk.

The first is by Hermann Weyl [W. 1938] in his review of Cartan's book [C. 1951].

> Nevertheless, I must admit that I found the book, like most of Cartan's papers, hard reading. Does the reason lie only in the great French geometric tradition on which Cartan draws, and the style and contents of which he more or less takes for granted as a common ground for all geometers, while we, born and educated in other countries, do not share it?

The second occurs in the middle of Singer and Sternberg's paper [Si–S. 1965].

> We now resume some of the principal formulae in coordinate notation with the idea of providing a partial guide to some of the writings of É. Cartan on the infinite groups and on the equivalence problem. We must confess that we find most of these papers extremely rough going and we can not follow all the arguments in detail.

Over the years, I have had the good fortune to be able to discuss the method of equivalence with many experts, in particular I would like to cite Chern, Kobayashi, Švec, Hermann, Vranceanu, Libermann, Kuranishi, Guillemin, Bryant, Shadwick, and Wilkens. They have all influenced my understanding of the method of equivalence.

The purpose of this work and the lecture series on which it is based is to describe the algorithm, showing, in particular, how it is applied to several pedagogical examples, and to a problem in control theory called state estimation of plants under feedback. To keep the prerequisites to a minimum, we have focused on problems in real geometry and ignored important examples from complex geometry (see [Ch. 1954] and [Ch. 1957]).

LECTURE 1

Equivalence Problems

The goal of the method of equivalence is to find necessary and sufficient conditions in order that "geometric objects" be "equivalent." The word *equivalent* here usually ends up meaning that the geometric objects are mapped onto each other by a class of diffeomorphisms characterized as the set of solutions of a system of differential equations. The necessary and sufficient conditions are found in the form of differential invariants of the geometric object under the class of diffeomorphisms. In this set of lectures we will restrict ourselves to classes of diffeomorphisms which can be described as solutions of a *first-order* system of differential equations or equivalently by conditions on their Jacobians. We will see that such problems are generally reducible to the following question.

THE EQUIVALENCE PROBLEM OF ÉLIE CARTAN. *Let* $\Omega_V = {}^t(\Omega_V^1, \cdots, \Omega_V^n)$ *be a coframe on an open set* $V \subset \mathbf{R}^n$, *and let* $\omega_U = {}^t(\omega_U^1, \cdots, \omega_U^n)$ *be a coframe on* $U \subset \mathbf{R}^n$, *and let* G *be a prescribed linear group in* $Gl(n, \mathbf{R})$, *then find necessary and sufficient conditions that there exists a diffeomorphism* $\Phi : U \longrightarrow V$ *such that for each* $u \in V$

$$\Phi^* \Omega_V|_{\Phi(u)} = \gamma_{VU}(u) \omega_U|_u, \tag{1}$$

where $\gamma_{VU}(u) \in G$.

(In the future we will always drop the base point notation and write (1) as $\Phi^*\Omega_V = \gamma_{VU}\omega_U$.)

Note that the conditions on Φ are expressed in terms of the Jacobian of Φ and hence correspond to first-order differential equations.

To get a feeling for the ubiquity of this problem let us look at some examples. Each of these examples will be treated further in these lectures and are chosen since each is among the simplest illustrating special points.

Example 1. *Invariants of Riemannian metrics under isometries.* In this case the class of diffeomorphisms is the full set of internal symmetries of the metric and the subject is the local theory of *Riemannian geometry.*

Thus we are given $(U, d\sigma^2)$ and $(V, d\Sigma^2)$ Riemannian metrics on open sets in $\mathbf{R}^n$ and want necessary and sufficient conditions that there exists a diffeomorphism $\Phi : U \longrightarrow V$ such that

$$\Phi^* d\Sigma^2 = d\sigma^2.$$

If we locally diagonalize the metrics

$$d\sigma^2 = \sum (\omega_U^i)^2, \qquad d\Sigma^2 = \sum (\Omega_V^i)^2,$$

then the problem is identical to finding necessary and sufficient conditions that there exists a diffeomorphism such that

$$\Phi^* \Omega_V = \gamma_{VU} \omega_U \quad \text{with } \gamma_{VU} \in O(n, \mathbf{R}),$$

where $O(n, \mathbf{R})$ is the orthogonal group.

We can express this formulation in a table.

OBJECTS		EQUIVALENCES
Riemannian metrics		isometries
	or	
basis of 1-forms adapted to the problem		diffeomorphisms which relative to the given basis have Jacobians in $O(n, \mathbf{R})$.

Example 2. *Invariants of a Riemannian metric under conformal transformations.* In this case the class of diffeomorphisms is larger than the internal symmetries of the metric and the subject is the local theory of *conformal geometry* (see [C. 1923]).

Thus we are given metrics as in Example 1, but we want diffeomorphisms such that

$$\Phi^* d\Sigma^2 = \lambda^2 d\sigma^2, \qquad \lambda \neq 0.$$

In terms of the coframe of Example 1,

$$\Phi^* \Omega_V = \gamma_{VU} \omega_U \quad \text{with } \gamma_{VU} \in CO(n, \mathbf{R}),$$

where $CO(n, \mathbf{R}) = \{\lambda S \mid \lambda \in \mathbf{R}^*, S \in SO(n, \mathbf{R})\}$.

We express this formulation in a table:

OBJECTS		EQUIVALENCES
Riemannian metrics		conformal transformations
	or	
basis of 1-forms adapted to the problem		diffeomorphisms which relative to the given basis have Jacobians in $CO(n, \mathbf{R})$.

Example 3. *Invariants of a first-order ordinary differential equation under diffeomorphisms of the form* $\Phi(x, y) = (\phi(x), \psi(y))$. In this case, the class of diffeomorphisms is a subset of the internal symmetries of the ordinary differential equation. This is called *Web geometry* in the plane (see [C. 1908], pp. 78-83).

Given (U, x, y) and (V, X, Y) open sets with coordinates in $\mathbf{R}^2$ and ordinary differential equations,

$$\frac{dy}{dx} = f(x, y) \text{ on } U, \quad \text{with } f \neq 0$$

and

$$\frac{dY}{dX} = F(X, Y) \text{ on } V, \text{ with } F \neq 0,$$

the usual symmetries of the ordinary differential equations are the diffeomorphisms which map integral curves into integral curves, that is, satisfy

$$\Phi^*(dY - F dX) = w(dy - f dx).$$

In the given problem we also have the condition on the Jacobian of the diffeomorphism that

$$\Phi^* \begin{pmatrix} dX \\ dY \end{pmatrix} = \begin{pmatrix} u & 0 \\ 0 & v \end{pmatrix} \begin{pmatrix} dx \\ dy \end{pmatrix}.$$

This is an overdetermined problem in that the conditions on the Jacobian of Φ are specified on

$$dX, \; dY, \; dY - F dX,$$

which form a linearly dependent set of 1-forms. If we note the relation

$$dY - F dX - dY + F dX = 0,$$

and if we change the forms on V by defining

$$\Omega_V^1 = F dX, \quad \Omega_V^2 = dY, \quad \Omega_V^3 = dY - F dX$$

then we have a constant coefficient relation

$$\Omega_V^1 - \Omega_V^2 + \Omega_V^3 = 0.$$

If we make the analogous choice of coframe on U, then an equivalence Φ will satisfy

$$\Phi^* \begin{pmatrix} \Omega_V^1 \\ \Omega_V^2 \\ \Omega_V^3 \end{pmatrix} = \begin{pmatrix} u & 0 & 0 \\ 0 & v & 0 \\ 0 & 0 & w \end{pmatrix} \begin{pmatrix} \omega_U^1 \\ \omega_U^2 \\ \omega_U^3 \end{pmatrix},$$

and hence will satisfy

$$0 = \Phi^*(\Omega_V^1 - \Omega_V^2 + \Omega_V^3) = u\omega_U^1 - v\omega_U^2 + w\omega_U^3,$$

and

$$\omega_U^1 - \omega_U^2 + \omega_U^3 = 0.$$

As a result

$$(u - v)\omega_U^2 - (w - u)\omega_U^3 = 0,$$

and we see that as long as $F \neq 0$, it necessarily follows that

$$u = v, \qquad w = u,$$

and an equivalence satisfies

$$\Phi^* \begin{pmatrix} \Omega_V^1 \\ \Omega_V^2 \end{pmatrix} = \begin{pmatrix} u & 0 \\ 0 & u \end{pmatrix} \begin{pmatrix} \omega_U^1 \\ \omega_U^2 \end{pmatrix}.$$

Conversely, given a diffeomorphism satisfying this last condition,

$$\begin{aligned} \Phi^*\Omega_V^3 &= \Phi^*(\Omega_V^2 - \Omega_V^1) = u\omega_U^2 - u\omega_U^1 \\ &= u(\omega_U^2 - \omega_U^1) = u\omega_U^3 \end{aligned}$$

which recovers the full conditions on the Jacobian.

We express the results in a table.

OBJECTS		EQUIVALENCES
first-order ordinary differential equations		diffeomorphisms which preserve integral curves of the form $\Phi(x,y) = (\phi(x), \psi(y))$
	or	
basis of 1-forms adapted to the problem		diffeomorphisms which relative to the given basis have Jacobians in the group of scalar matrices

Example 4. *Invariants of a second-order ordinary differential equation under diffeomorphisms of the form* $\Phi(x,y) = (x, \phi(y))$. In this case the class of diffeomorphisms is a subset of the internal symmetries of the ordinary differential equations. This is called the *geometry of time-fixed Newton's equations* (see [C. 1937]).

Given (U, x, y, y') and (V, X, Y, Y') open sets with standard jet coordinates on the 1-jets of mappings of the line into the line (see [G. 1983]) and second-order ordinary differential equations

$$y'' = f(x, y, y'), \qquad Y'' = F(X, Y, Y'),$$

the usual symmetries of the second-order ordinary differential equations are the diffeomorphisms which map integral curves into integral curves, that is,

$$\Phi^* \begin{pmatrix} dY - Y'dX \\ dY' - FdX \end{pmatrix} = \begin{pmatrix} v & 0 \\ z & w \end{pmatrix} \begin{pmatrix} dy - y'dx \\ dy' - fdx \end{pmatrix}$$

Exercise. Why is the zero in the upper right corner?

We also have the conditions on the Jacobians of the diffeomorphisms

$$\Phi^* \begin{pmatrix} dX \\ dY \end{pmatrix} = \begin{pmatrix} 1 & 0 \\ 0 & u \end{pmatrix} \begin{pmatrix} dx \\ dy \end{pmatrix}.$$

As before, this is an overdetermined problem on generators

$$dX, \;\; dY, \;\; dY - Y'dX, \;\; dY' - FdX$$

and we can proceed similarly, but there is a twist. The relation

$$(dY - Y'dX) - dY + Y'dX = 0$$

would seem to suggest modifying the forms to make one of them $Y'dX$, but dX is obviously invariant under the admissible diffeomorphisms, and we would not like to lose this fact, hence define

$$\Omega_V^1 = dX, \quad \Omega_V^2 = \frac{dY}{Y'}, \quad \Omega_V^3 = dY' - F dX, \quad \Omega_V^4 = \frac{(dY - Y'dX)}{Y'},$$

in order to get a constant coefficient relation

$$\Omega_V^4 - \Omega_V^2 + \Omega_V^1 = 0.$$

Now applying the same sort of argument as in Example 3, we find that the original problem is identical to finding diffeomorphisms that satisfy the Jacobian condition

$$\Phi^* \begin{pmatrix} \Omega_V^1 \\ \Omega_V^2 \\ \Omega_V^3 \end{pmatrix} = \begin{pmatrix} 1 & 0 & 0 \\ 0 & 1 & 0 \\ -z & z & w \end{pmatrix} \begin{pmatrix} \omega_U^1 \\ \omega_U^2 \\ \omega_U^3 \end{pmatrix}.$$

Here, of course, the lowercase omegas are defined analogously to the uppercase omegas.

Exercise. Verify the Jacobian condition.

It is this sort of special observation that caused the difficulty in understanding Cartan's examples well enough to solve new problems. We will see in Lecture 5 that there is a general algorithmic technique for reducing an overdetermined problem to a determined equivalence problem, and that this algorithm predicts the above change of frame.

We express these results in a table.

OBJECTS		EQUIVALENCES
second-order ordinary differential equations		diffeomorphisms which preserve integral curves of the form $\Phi(x, y) = (x, \phi(y))$
	or	
basis of 1-forms adapted to the problem		diffeomorphisms which relative to the given basis have Jacobians in a given two parameter non-Abelian group.

Example 5. *Invariants of third-order ordinary differential equations under contact transformations.* In this case the class of diffeomorphisms is the full set of internal symmetries. This is called the *contact geometry of a third-order ordinary differential equation* (see [Ch. 1940]). Given (U, x, y, y', y'') and (V, X, Y, Y', Y'') open sets with standard jet coordinates on the 2-jets of mappings of the line into the line and third-order ordinary differential equations

$$y''' = f(x,y,y',y''), \qquad Y''' = F(X,Y,Y',Y''),$$

the symmetries of these third-order ordinary differential equations are, as usual, the diffeomorphisms of the plane which preserve integral curves. If we define

$$\Omega^1 = dY - Y'dX, \quad \Omega^2 = dY' - Y''dX, \quad \Omega^3 = dY'' - FdX, \quad \Omega^4 = dX,$$

then the problem is identical to finding diffeomorphisms satisfying

$$\Phi^* \begin{pmatrix} \Omega_V^1 \\ \Omega_V^2 \\ \Omega_V^3 \\ \Omega_V^4 \end{pmatrix} = \begin{pmatrix} a & 0 & 0 & 0 \\ b & c & 0 & 0 \\ e & k & g & 0 \\ h & i & 0 & j \end{pmatrix} \begin{pmatrix} \omega_U^1 \\ \omega_U^2 \\ \omega_U^3 \\ \omega_U^4 \end{pmatrix}.$$

Exercise. The fact that we are restricted to contact transformations is reflected in the presence of the zero in the last row of the matrix. Show why this is true, and how it would differ if we considered point transformations.

We express these results in a table.

OBJECTS		EQUIVALENCES
third-order ordinary differential equations		contact transformations
	or	
basis of 1-forms adapted to the problem		diffeomorphisms which, relative to the given basis, have Jacobians in a given nine parameter non-Abelian group

Example 6. *Invariants of particle or field Lagrangians under their "natural symmetries."* In this case we have to figure out how to characterize the class of diffeomorphisms we want to call "natural symmetries." This is called the *geometry of the integral*, and was classically written as the geometry of $\delta \int L\, dA$ (see [C. 1930]).

A first-order Lagrangian is a function on the 1-jets

$$L : J^1(\mathbf{R}^m, \mathbf{R}^n) \longrightarrow \mathbf{R}.$$

Given

$$(U, x^1, \cdots, x^m, z^1, \cdots, z^n, p_1^1, \cdots, p_m^n) = (U, x, z, p)$$

and

$$(V, X^1, \cdots, X^m, Z^1, \cdots, Z^n, P_{11}, \cdots, P_{mn}) = (V, X, Z, P),$$

open sets with standard jet coordinates on the 1-jets of mappings of the m-space into n-space (see [G-S. 1985]), and two Lagrangians

$$l(x,z,p), \qquad L(X,Z,P),$$

we want to characterize those diffeomorphisms $\Phi : U \longrightarrow V$ that satisfy

$$\int_{j^1(\alpha)} \Phi^*(L\, dX^1 \wedge \cdots \wedge dX^m) = \int_{j^1(\alpha)} l\, dx^1 \wedge \cdots \wedge dx^m$$

for all 1-graphs $j^1(\alpha)^*$ of maps $\alpha : \mathbf{R}^m \longrightarrow \mathbf{R}^n$.

By using a cutoff function argument, this implies

$$j^1(\alpha)^*(\Phi^*\Psi - \psi) = 0,$$

where

$$\Psi = L\, dX^1 \wedge \cdots \wedge dX^m, \qquad \psi = l\, dx^1 \wedge \cdots \wedge dx^m.$$

The characterization of the contact ideal $\mathcal{I}$ (see [P-R-S. 1979]) then implies

$$\Phi^*\Psi \equiv \psi \mod \mathcal{I}.$$

In the case $m = 1$, the case of particle Lagrangians, this condition becomes

$$\Phi^*\Psi = \psi + \sum b_\alpha(dz^\alpha - p^\alpha dx).$$

Unfortunately, this last condition does not define an equivalence relation, because diffeomorphisms Φ need not carry 1-graphs to 1-graphs, but it follows that transitivity is equivalent to the diffeomorphism being a contact transformation; that is, given

$$\Theta_V^\alpha = dZ^\alpha - \sum P^\alpha dX, \qquad \theta_U^\alpha = dz^\alpha - \sum p^\alpha dx,$$

then an equivalence Φ satisfies

$$\Phi^*\Theta_V^\alpha \in \{\theta_U^\alpha\}.$$

Exercise. Use the identity diffeomorphism, and an arbitrary diffeomorphism to see how to verify this claim for particle Lagrangians. (The case of field Lagrangians requires technical results about the contact ideal.)

To state these conditions in terms of the Jacobian acting on a basis of 1-forms we take the Lagrangian m-forms Ψ and ψ and factor them arbitrarily as monomials, that is, write them

$$\Psi = \Omega_V^1 \wedge \cdots \wedge \Omega_V^m, \qquad \psi = \omega_U^1 \wedge \cdots \wedge \omega_U^m.$$

As such, the $\{\Omega_V^i\}$ and $\{\omega_U^i\}$ are defined up to an action of $Sl(m, \mathbf{R})$, the special linear group. This of course assumes the Lagrangians are nowhere zero.

The conditions now only involve

$$\Omega_V = {}^t(\Omega_V^1, \cdots, \Omega_V^m) \text{ and } \Theta_V = {}^t(\Theta_V^1, \cdots, \Theta_V^m)$$

and their lowercase analogues. Hence this is an underdetermined problem, and we extend the forms to a coframe arbitrarily. In order to be specific take

$$\mathbf{H}_V = {}^t(dP_1^1, \cdots, dP_n^m), \qquad \eta_U = {}^t(dp_1^1, \cdots, dp_n^m).$$

Now collecting our conditions, we see that the "natural equivalences" satisfy the conditions

$$\Phi^* \begin{pmatrix} \Theta_V \\ \Omega_V \\ \mathbf{H}_V \end{pmatrix} = \begin{pmatrix} A_1 & 0 & 0 \\ b_1 & D & 0 \\ B & b_2 & A_2 \end{pmatrix} \begin{pmatrix} \theta_U \\ \omega_U \\ \eta_U \end{pmatrix}$$

with $D \in Sl(n, \mathbf{R})$, $A_1 \in GL(n, \mathbf{R})$, $A_2 \in GL(mn, \mathbf{R})$, and b_1, b_2, B arbitrary matrices of the appropriate sizes.

We express these results in a table.

OBJECTS		EQUIVALENCES
first-order Lagrangians		contact transformations which preserve the associated integral functional
	or	
basis of 1-forms adapted to the problem		diffeomorphisms which relative to the given basis have Jacobians in a specified group.

Example 7. *Invariants of underdetermined systems of ordinary differential equations under diffeomorphisms of the form*

$$\Phi(x, u) = (\phi(x), \psi(x, u)), \quad \textit{where } x \in \mathbf{R}^m, \quad u \in \mathbf{R}^p.$$

This is called the *geometry of control systems under feedback.*

Given (V, X, U) and (U, x, u) open sets with coordinates in $\mathbf{R}^m \times \mathbf{R}^n$ and an underdetermined system of ordinary differential equations

$$\frac{dx}{dt} = f(x, u), \qquad \frac{dX}{dt} = F(X, U),$$

the above problem is an overdetermined problem. When we discuss the overdetermined algorithm in Lecture 5, we will see that the choice of coframe

$$\begin{aligned} &\mathbf{H}_V = A_0 dX && \eta_U = a_0 dx \\ &\mathbf{M}_V = dU && \mu_U = du \\ &A_0 F = {}^t(1, 0, \cdots, 0) && a_0 f = {}^t(1, 0, \cdots, 0), \end{aligned}$$

where A is a nonsingular matrix satisfying $A_0 F = {}^t(1, 0, \cdots, 0)$, and similarly a_0 is a nonsingular matrix satisfying $a_0 f = {}^t(1, 0, \cdots, 0)$, leads to the characterization of diffeomorphisms in our original problem in the form

$$\Phi^* \begin{pmatrix} \mathbf{H} \\ \mathbf{M} \end{pmatrix} = \begin{pmatrix} A & 0 \\ B & C \end{pmatrix} \begin{pmatrix} \eta \\ \mu \end{pmatrix},$$

where the matrix A has the form

$$A = \begin{pmatrix} 1 & * \\ 0 & ** \end{pmatrix}.$$

We can express this formulation in a table.

OBJECTS		EQUIVALENCES
control systems		feedback transformations
	or	
basis of 1-forms adapted to the problem		diffeomorphisms which, relative to the given basis, have Jacobians in a given group.

This example will be one of the main focuses of this book and in particular will be the topics of Lectures 8 and 9.

This should give enough examples of diverse sorts to indicate that the equivalence problem is one of utmost importance in mathematics. Now my job is to describe how to solve such problems and your job is to find and solve new problems.

LECTURE 2

Lifting of Equivalence Problems to G-Spaces

Given $\Omega_V = {}^t(\Omega_V^1, \cdots, \Omega_V^n)$ and $\omega_U = {}^t(\omega_U^1, \cdots, \omega_U^n)$ coframes on open sets U and V, and a connected linear group $G \subset GL(n, R)$, we want to study the existence of diffeomorphisms $\Phi : U \longrightarrow V$ satisfying $\Phi^* \Omega_V = \gamma_{VU} \omega_U$.

The first idea is to build the group action into the question in a natural way. This is achieved by lifting the problem to the associated spaces $U \times G$, and $V \times G$ with the natural left actions, that is,

$$C(p, S) = (p, CS), \qquad C, S \in G, \quad p \in U \text{ or } V.$$

In general, a space on which G acts on the left is called a G-space. Those familiar with fiber bundles should replace our product G-spaces by principal left G-bundles.

Let $\Pi_V : V \times G \longrightarrow V$ and $\Pi_U : U \times G \longrightarrow U$ be the natural projection and define new column vectors of 1-forms on $V \times G$ and $U \times G$ by

$$\Omega|_{(V,T)} = T \pi_V^* \Omega_V, \qquad \omega|_{(U,S)} = S \pi_U^* \omega_U.$$

The following simple result is the key to the usefulness of the lifting procedure.

Proposition. *There exists a diffeomorphism $\Phi : U \longrightarrow V$ satisfying*

$$\Phi^* \Omega_V = \gamma_{VU} \omega_U \quad \textit{with } \gamma_{VU} : U \longrightarrow G$$

if and only if there exists a diffeomorphism $\Phi^1 : U \times G \longrightarrow V \times G$ such that $\Phi^{1*} \Omega = \omega$.

Although the proof is easy, it will set the stage to look at it, especially since this diffeomorphism Φ^1 covers Φ, i.e., the diagram with the natural projections

$$\begin{array}{ccc} U \times G & \xrightarrow{\Phi^1} & V \times G \\ {\scriptstyle \pi_U} \downarrow & & {\scriptstyle \pi_V} \downarrow \\ U & \xrightarrow{\Phi} & V \end{array}$$

commutes. Further, Φ^1 is unique and automatically satisfies $\Phi^1(u, CS) = C\Phi^1(u, S)$.

Proof. Let $\Phi^1 : U \times G \longrightarrow V \times G$ be defined by

$$\Phi^1(u, S) = (\Phi(u, S), T(u, S))$$

with $\Phi(u, S) \in V$ and $T(u, S) \in G$. The condition $\Phi^{1*}\Omega = \omega$ means

$$\Phi^{1*} T\pi_V^* \Omega_V = S\pi_U^* \omega_U$$

or

$$(\pi_V \circ \Phi^1)^* \Omega_V = (T \circ \Phi^1)^{-1} S\pi_U^* \omega_U$$

or

$$\Phi^* \Omega_V = T(u, S)^{-1} S\pi_U^* \omega_U.$$

Since the Ω_V contains a basis for the differentials on V and the ω_U contains a basis for the differentials on U, the partial derivatives of Φ with respect to the group variables must vanish and, under the assumption that G is connected, $\Phi(u, S) = \Phi(u)$. In particular this implies

$$T(u, S)^{-1} S = \gamma_{VU}(u)$$

and completes the equivalence. Note that this last equation implies

$$T(u, S) = S\gamma_{VU}^{-1}(u)$$

and hence the map Φ^1 has the form

$$\Phi^1(u, S) = (\Phi(u), S\gamma_{VU}^{-1}(u)). \tag{1}$$

As such, given an equivalence $\Phi : U \longrightarrow V$ we define Φ^1 by (1), which works since

$$\begin{aligned} \Phi^{1*} \Omega|_{(\Phi(u), S\gamma_{VU}^{-1})} &= S\gamma_{VU}^{-1} \Phi^* \Omega_V \\ &= S\gamma_{VU}^{-1} \gamma_{VU} \omega_U = S\omega_U = \omega. \end{aligned}$$

It now follows from the explicit representation of the diffeomorphism Φ^1 that

$$\pi_V \circ \Phi^1 = \Phi \circ \pi_U, \qquad \Phi^1(u, CS) = C\Phi^1(u, S)$$

so that the natural equivariance that might be desired is automatic.

Note that, in the example of Riemannian geometry, the method tells us to lift the problem to the space $U \times O(n, \mathbf{R})$. The construction of the lifts and the γ_{VU} are the ingredients of the bundle of orthonormal frames, the natural arena for the modern treatment of this geometry.

Similarly, the example of conformal geometry constructs the natural arena for its modern treatment, but in this case there is more to come.

In practice it is often true that some problems are underdetermined in the sense that conditions on the diffeomorphisms do not involve the whole coframe. We will call such a problem *underdetermined* and note that it can always be replaced by a determined problem as we did in Examples 5 and 6. We simply extend the partial coframe to a coframe and extend the conditions on the extra forms to be as general as possible.

Example 8. *Invariants of a Pfaffian equation or, equivalently, a total differential equation.* In this case, the class of all diffeomorphisms is the full set of internal symmetries.

Thus we are given two Pfaffian equations

$$\Omega^1_V = 0 \quad \text{on a neighborhood } V, \qquad \omega^1_U = 0 \quad \text{on a neighborhood } U.$$

We arbitrarily extend Ω^1 and ω^1 to coframes, say $\{\Omega^i_V\}$, and $\{\omega^i_U\}$, and extend the condition that a diffeomorphism Φ preserve the equations

$$\Phi^*\Omega^1_V = w\ \omega^1_U$$

to

$$\Phi^*\begin{pmatrix}\Omega^1_V\\ \Omega^2_V\\ \vdots\\ \Omega^n_V\end{pmatrix} = \begin{pmatrix} w & 0 & \cdots & 0\\ a_{21} & a_{22} & \cdots & a_{2n}\\ \vdots & \vdots & \ddots & \vdots\\ a_{n1} & a_{n2} & \cdots & a_{nn}\end{pmatrix}\begin{pmatrix}\omega^1_U\\ \omega^2_U\\ \vdots\\ \omega^n_U\end{pmatrix}.$$

Thus in this case the group is

$$G = \left(\begin{array}{c|c} w & 0\\ \hline * & **\end{array}\right).$$

In the last lecture we saw two examples of filling out coframes arbitrarily to capture information on partial frames. To understand the statement "extend the conditions on the extra forms to be as general as possible," let us look again at the third-order ordinary differential equation.

Here we extended

$$\begin{aligned} \Omega^1 &= dY - Y'dX, & \omega^1 &= dy - y'dx\\ \Omega^2 &= dY' - Y''dX, & \omega^2 &= dy' - y''dx\\ \Omega^3 &= dY'' - FdX, & \omega^3 &= dy'' - fdx \end{aligned}$$

by

$$\Omega^4 = dX, \qquad \omega^4 = dx$$

and noted that a contact transformation could only pull back f^*dX to a linear combination of dx, dy, and dy'. Thus, this is part of the solution of the exercise in the first lecture to explain the zero in the last row of the Jacobian of an equivalence for this problem.

Maurer–Cartan Forms and the dSS^{-1} Yoga.

We now need to take an excursion into Lie theory, in particular the theory of Maurer–Cartan forms. Thus let G be a Lie group and let R_C denote right multiplication by $C \in G$. If we choose a basis $\{\omega^i|_e\}$ of T_e^*, the cotangent space at the identity e, then we may define global differential forms by

$$\omega^i|_A = R^*_{A^{-1}}(\omega^i|_e) \quad \text{for } A \in G.$$

Since

$$R^*_C(\omega^i|_{AC}) = R^*_C \circ R^*_{(AC)^{-1}}(\omega|_e) = R^*_C \circ R^*_{C^{-1}} \circ R^*_{A^{-1}}(\omega|_e) = \omega^i|_A,$$

these are a basis for the right invariant 1-forms on the group, and are called a set of Maurer–Cartan forms.

Matters being so, a set of right invariant Maurer–Cartan forms $\{\omega^i\}$ define functions C^i_{jk} via the equations

$$d\omega^i = \frac{1}{2}\sum C^i_{jk}\,\omega^j \wedge \omega^k,$$

where we assume $C^i_{jk} = -C^i_{kj}$. The right translational invariance immediately implies that

$$C^i_{jk}(AC) = C^i_{jk}(A), \tag{2}$$

and hence choosing $C = A^{-1}$, we see that the functions C^i_{jk} are in fact constants. This fact is known as the first fundamental theorem of Lie. These constants are called the structure constants of G relative to the choice of Maurer–Cartan forms.

Exercise. Verify the above functional relations (2) on C^i_{jk}.

There is a basis-free approach to the information contained in the Maurer–Cartan forms, which explains why they are naturally packaged in a matrix in the case of linear groups. This is the theory of the *T_e-valued Maurer–Cartan form.*

Let I be the image of the identity homomorphism of the tangent space at the identity, under the isomorphism $\mathrm{Hom}(T_e, T_e) \approx T_e \otimes T_e^*$, and define

$$\varpi(S) = 1 \otimes R^*_{S^{-1}}(I) \in T_e \otimes T^*_S(G).$$

This $\varpi(S)$ is called *the right invariant T_e-valued Maurer–Cartan form.*

Exercise. Verify that $1 \otimes R^*_C \varpi(SC) = \varpi(S)$.

Now assume that $G \subset Gl(n, \mathbf{R})$, that is, G is a linear group. Then

$$T_eG \subset T_eGl(n, \mathbf{R}),$$

and since $Gl(n, \mathbf{R})$ is an open submanifold of the vector space of all matrices $M(n, \mathbf{R})$,

$$T_eGl(n, \mathbf{R}) \approx M(n, \mathbf{R})$$

under the canonical isomorphism sending the equivalence class of curves $[e+tA]$ defining a tangent vector in $T_eGL(n,\mathbf{R})$ onto the matrix A.

We now compute $\varpi(S)$ for $S \in Gl(n,\mathbf{R})$, where we view

$$\varpi(S) \in M(n,\mathbf{R}) \otimes T_S^*.$$

The standard projections on the $(i,\ j)$ entry

$$\pi_{ij} : Gl(n,\mathbf{R}) \longrightarrow \mathbf{R},$$

give a global coordinate system. We will write $\pi_{ij}(S) = S_{ij}$.

Let E_{ik} be the matrix with 1 in the $(i,\ k)$ entry and zeros in the rest, then, since

$$\langle [e+tE_{ik}], d\pi_{jl}\rangle = \frac{d}{dt}\,\pi_{jl}(e+tE_{ik})|_{t=0} = \delta_{ji}\delta_{lk},$$

we see that $\{E_{ik}\}$ is a dual basis to $\{d\pi_{jl}\}$.

As such

$$\varpi(e) = \sum E_{ik} \otimes d\pi_{ik} = (d\pi_{ik}).$$

Exercise. Verify this representation of $\varpi(e)$.

Let us note

$$\pi_{ij} \circ R_C(S) = \pi_{ij}(SC) = \sum \pi_{ik}(S)\pi_{kj}(C),$$

and hence

$$R_C^* d\pi_{ij}|_{SC} = \sum d\pi_{ik}|_S\, \pi_{kj}(C).$$

If we introduce the shorthand notation

$$dS = (d\pi_{ij}|_S),$$

we have shown $R_C^* d(SC) = dS\,C$.

This implies that

$$R_C^* d(SC)\,(SC)^{-1} = dS\,CC^{-1}S^{-1} = dS\,S^{-1},$$

and hence dSS^{-1} is a $M(n,\mathbf{R})$-valued right invariant Maurer–Cartan form. At $S = e$

$$dee^{-1} = (d\pi_{ij}) = \varpi(e),$$

and by right invariance

$$\boxed{\varpi(S) = dSS^{-1}.}$$

If we are given a Lie subgroup $j : G \longrightarrow Gl(n,\mathbf{R})$ we find a set of Maurer–Cartan forms by finding a maximal independent set of entries of $1 \otimes j^*\varpi(S)$.

The exterior derivative

$$d\varpi(S) = d(dSS^{-1}) = -dS \wedge d(S^{-1}) = dSS^{-1} \wedge dSS^{-1} = \varpi(S) \wedge \varpi(S)$$

gives *the Maurer–Cartan equations.*

The Maurer–Cartan forms encode the data of the Lie algebra of the group and hence the local structure of the group. They are also one of the key tools needed to describe global properties of the group. For example, let $j : H \longrightarrow G$ be a Lie subgroup of a connected Lie group G, and let $\dim H = m$ and $\dim G = r$; then we may choose a basis of $T_e^* G$

$$\{\omega^1, \cdots, \omega^m, \omega^{m+1}, \cdots, \omega^r\}$$

with

$$\{\omega^{m+1}, \cdots, \omega^r\} \subset \ker j^*,$$

and use this basis to define a set of Maurer–Cartan forms adapted to the subgroup H. Since

$$0 = j^* d\omega^a|_e = \sum C^a_{\alpha\beta}\; j^*\omega^\alpha|_e \wedge j^*\omega^\beta|_e$$

$$1 \leq \alpha, \quad \beta \leq m, \qquad m+1 \leq a \leq r$$

we see

$$C^a_{\alpha\beta} = 0,$$

and hence that

$$\{\omega^{m+1}, \cdots, \omega^r\}$$

is a completely integrable system.

If we let $A \in H$, then $j \circ R_{A^{-1}} = R_{A^{-1}} \circ j$. Thus, H is an integral manifold of maximal dimension of $\{\omega^{m+1}, \cdots, \omega^r\}$ through e. Since this differential system is right invariant, each right coset of H is also an integral manifold. Finally, since there is a connected component of a right coset through each point of G and it is a maximal integral manifold by right invariance, we see by the uniqueness part of the Frobenius theorem that these are the maximal connected integral manifolds.

The following theorem is an example of the power of these ideas.

THEOREM. *Let $f_i : M \longrightarrow G$ for $i = 1, 2$ be two immersions of a connected manifold M into a Lie group G, then necessary and sufficient conditions that there exists an element $S \in G$, such that*

$$R_S \circ f_1 = f_2,$$

are that a set of Maurer–Cartan forms $\{\omega^i\}$ satisfy

$$f_1^*\omega^i = f_2^*\omega^i.$$

Proof. *Let $j : \Delta \longrightarrow G \times G$ define the diagonal subgroup*

$$\Delta = \{(S, S)|\; S \in G\}.$$

Letting

$$\pi_L : G \times G \longrightarrow G, \qquad \pi_R : G \times G \longrightarrow G$$

be the left and right projections, we see that

$$\{\pi_L^*\omega^i - \pi_R^*\omega^i\}$$

is a differential system of rank r which lies in the kernel of j^* at e. Hence this is the differential system constructed from the diagonal subgroup Δ.

By hypothesis the map $h : M \longrightarrow G \times G$ defined by

$$h(x) = (f_1(x), f_2(x))$$

satisfies

$$\begin{aligned} h^*(\pi_L^*\omega^i - \pi_R^*\omega^i) &= (\pi_L \circ h)^*\omega^i - (\pi_R \circ h)^*\omega^i \\ &= f_1^*\omega^i - f_2^*\omega^i = 0, \end{aligned}$$

and hence $h : M \longrightarrow G \times G$ is an integral manifold of the differential system whose maximal integral manifolds are the connected components of the right cosets of Δ. As such, there exists $(A, B) \in G \times G$ such that

$$(f_1(x), f_2(x)) \in \Delta(A, B) \quad \text{for all } x \in M.$$

Equivalently

$$f_1(x)A^{-1} = f_2(x)B^{-1},$$

and hence defining $S = A^{-1}B$, we have $R_S \circ f_1 = f_2$, as required.

The following is an important special case.

COROLLARY. *Let G be a connected Lie group; then a diffeomorphism $f : G \longrightarrow G$ is a right translation by an element $C \in G$ if and only if a set of Maurer–Cartan forms ω_i satisfies*

$$f^*\omega_i = \omega_i.$$

Proof. Let $f(e) = C$ and let $f_1 = f$ and $f_2 = R_C$ in the last theorem; the result then follows.

This corollary characterizes which diffeomorphisms are right translations.

Now I want to discuss one final result from Lie theory, known as *the local converse to the Third Lie Theorem.*

THEOREM. *Let C^i_{jk}, $1 \le i, j, k \le n$, be constants satisfying*

$$C^i_{jk} = -C^i_{kj},$$

$$\sum_j C^i_{jr}C^j_{sk} + C^i_{js}C^j_{kr} - C^i_{jk}C^j_{rs} = 0,$$

then there exist linearly independent 1-forms $\omega^1, \cdots, \omega^n$ in a neighborhood of the origin in $\mathbf{R}^n$ satisfying

$$d\omega^i = \frac{1}{2}\sum C^i_{jk}\omega^j \wedge \omega^k.$$

This result is the key to establishing the existence of many interesting examples of Pfaffian systems, as well as being the first step in the form-theoretical approach to the construction of Lie groups which will be taken up in Lectures 6 and 7. As such, I want to share my viewpoint on this theorem.

Note first that the indicial horror listed as the second set of conditions in the theorem simply consists of the necessary conditions needed to solve

$$d\omega^i = \frac{1}{2}\sum C^i_{jk}\omega^j \wedge \omega^k,$$

given by $d(d\omega^i) = 0$.

Next, I note that if there is a particular solution $\{\omega^i_0\}$ that is linearly independent, then the most general solution, dropping the condition of linear independence, is $\{f^*\omega^i_0\}$, where $f : \mathbf{R}^n \longrightarrow \mathbf{R}^n$ is an arbitrary differentiable map.

The fact that the $\{f^*\omega^i_0\}$ are solutions is immediate; the fact that every other solution could be represented in this way is a consequence of the technique of the graph, which will be discussed in Lecture 6.

An obvious solution, ignoring the linear independence, is $\omega^i = 0$. Since there is such a rich potential set of solutions, a reasonable idea is to try to construct a homotopy through solutions from this trivial solution, to a linearly independent set. I will now sketch the proof following Cartan's original ideas (see [C. 1951] or [F. 1963]). Let $(a^1, \cdots, a^n, t)$ be coordinates on $\mathbf{R}^n \times \mathbf{R}$, then in order that 1-forms

$$\beta^i = \sum h^i_r da^r \quad \text{defined on } \mathbf{R}^n \times \mathbf{R},$$

be such a homotopy, we need

$$\frac{d}{dt}\, h^i_r = \delta^i_r + \sum C^i_{jk} a^j h^k_r \quad \text{with } h^i_r(a, 0) = 0,$$

a system of ordinary differential equations with parameters $a = (a^1, \cdots, a^n)$. The claim is that

$$\omega^i|_a = \beta^i|_{(a,1)}$$

satisfies the theorem.

Since $h^i_r(0, t) = t\delta^i_r$ it follows that $\omega^1 \wedge \cdots \wedge \omega^n \neq 0$, in a neighborhood of the origin in $\mathbf{R}^n$.

Next we show that the 2-forms

$$\theta^i = d_{\mathbf{R}^n}\beta^i - \frac{1}{2}\sum C^i_{jk}\beta^j \wedge \beta^k$$

are identically zero on $\mathbf{R}^n \times \mathbf{R}$, and hence that the desired properties hold for $t = 1$. The idea in verifying this identity is to show that

$$\frac{d}{dt}\theta^i = \sum C^i_{jk} a^j \theta^k$$

with the initial conditions $\theta^i(a, 0) = 0$; hence by uniqueness of solutions to ordinary differential equations we have $\theta^i(a, t) = 0$.

LECTURE 3

The Structure Equations

Returning to the lifted equivalence problem, we will analyze the lifted differential forms

$$\omega = S\omega_U \quad \text{on } U \times G,$$

with the understanding that every construction will be repeated for the analogous forms on $V \times G$. *In practice, in fact, we usually want to find the invariants of a geometric structure under given equivalences, and these can be determined without introducing the analogous forms on $V \times G$.* The reason for wording the equivalence problem as I have is that it encompasses both problems and makes it easier to see the ideas.

On $U \times G$ we can differentiate the lifted forms to find

$$\begin{aligned} d\omega &= dS \wedge \omega_U + S\, d\omega_U \\ &= dSS^{-1} \wedge S\omega_U + Sd\omega_U. \end{aligned}$$

The matrix dSS^{-1} is, of course, the Maurer–Cartan matrix of right invariant forms on G, hence

$$(dSS^{-1})^i_j = \sum a^i_{j\rho}\, \pi^\rho,$$

where π^ρ is a basis for the Maurer–Cartan forms and the $a^i_{j\rho}$ are constants.

Exercise. Why are the coefficients $a^i_{j\rho}$ constant?

It is helpful in reading Cartan to know that the infinitesimal generators of the linear action $V \times G \longrightarrow V$ obtained from the basis $\{\varepsilon_\rho\}$ dual to the basis $\{\pi^\rho\}$ are given by vector fields $\{X_\rho\}$, where in the usual vector space coordinates x^i

$$X_\rho = \sum a^i_{j\rho} x^j \frac{\partial}{\partial x^i}.$$

The second Lie theorem then asserts

$$[X_\rho, X_\sigma] = \sum C^\alpha_{\rho\sigma} X_\alpha.$$

Recalling that the forms ω_U are basic, that is, both coefficients and differentials can be expressed in terms of coordinates on U alone, we can write the exterior derivatives in the group-fiber representation

$$d\omega^i = \sum a^i_{j\rho}\, \pi^\rho \wedge \omega^j + \frac{1}{2} \sum \gamma^i_{jk}(u, S)\, \omega^j \wedge \omega^k. \tag{3}$$

Note that the coefficients of the term which is quadratic in the $\{\omega^i\}$ has G dependence since three actions of the factor S had to be absorbed into those coefficients.

In practice it is time-consuming to write the exterior derivatives in the above form. Therefore the gameplan is to find a way to utilize the information in this group theoretical decomposition, but to do it with minimal computation.

Thus let us write the exterior derivatives in the following form,

$$d\omega^i = \sum \Delta^i_j \wedge \omega^j,$$

where no assumption is made on the Δ^i_j.

If we now subtract the group-fiber representation (3) from the above representation we find

$$0 = \sum(\Delta^i_k - a^i_{kp}\pi^p) \wedge \omega^k - \frac{1}{2}\sum \gamma^i_{jk}\omega^j \wedge \omega^k.$$

Part of the information in this last equation can be extracted by the use of the ubiquitous Cartan's lemma.

CARTAN'S LEMMA. *Let $\{\omega^i\}$ be an independent set of 1-forms, and let $\{\pi^i\}$ be an arbitrary set of 1-forms of the same finite cardinality; then*

$$\sum \pi_i \wedge \omega^i = 0$$

holds if and only if

$$\pi_i = \sum C_{ij}\omega^j, \quad \textit{where } C_{ij} \textit{ is a symmetric matrix.}$$

Exercise. Prove Cartan's lemma.

A consequence of Cartan's lemma applied to our last equation is that

$$\Delta^i_j - \sum a^i_{jp}\pi^p \equiv 0 \mod (\omega^1, \cdots, \omega^n).$$

We will often have congruences $\mod (\omega^1, \cdots, \omega^n)$ and will shorten this expression by saying "mod base." Thus

$$\boxed{\text{mod base} = \text{mod } (\omega^1, \cdots, \omega^n).}$$

Now if we have any constant coefficient relation on the matrix of Maurer–Cartan forms, say

$$\sum b^j_i a^i_{jp}\pi^p = 0,$$

we see

$$\sum b^j_i \Delta^i_j \equiv 0 \text{ mod base}.$$

The 1-forms $\sum b^j_i \Delta^i_j$, where b^j_i ranges over a set of defining relations, are called *principal components of order* 1.

The idea is to locate principal components of order 1, *and if possible use this observation to modify the matrix* Δ^i_j *to satisfy the defining relations on the Maurer–Cartan form matrix.* Since the defining relations on the Maurer–Cartan matrix are in fact the defining relations on the Lie algebra of G, we have described an algorithmic construction of a representation of the exterior derivatives of ω in the form

$$d\omega^i = \sum \pi^i_k \wedge \omega^k + \frac{1}{2}\sum \gamma^i_{jk}\omega^j \wedge \omega^k,$$

where the matrix $(\pi^i_j) \equiv (\sum a^i_{j\rho}\pi^\rho)$ mod base is now a Lie algebra valued differential form. The terms involving the coefficients γ^i_{jk} are called *torsion terms*, and the coefficients themselves are called the *torsion coefficients.*

These equations do not define the torsion coefficients nor the 1-forms π^ρ uniquely. *The next idea is to use this ambiguity to try to simplify, even eliminate if possible, the torsion coefficients by modifying* π^i_j *by multiples of* ω^k *in such a way that it is still Lie algebra valued.* This process is called *Lie algebra valued compatible absorption.* It will then still be true that the π^i_j are Maurer–Cartan forms mod base.

To understand the ambiguity, let us look at two representations of the exterior derivatives, the original group-fiber representation (3)

$$d\omega^i = \sum a^i_{k\rho}\pi^\rho \wedge \omega^k + \frac{1}{2}\sum \gamma^i_{jk}\omega^j \wedge \omega^k$$

and another representation

$$d\omega^i = \sum a^i_{k\rho}\varpi^\rho \wedge \omega^k + \frac{1}{2}\sum \Gamma^i_{jk}\omega^j \wedge \omega^k,$$

where a priori nothing is assumed about ϖ^ρ or Γ^i_{jk}. Subtracting these representations gives

$$0 = \sum a^i_{k\rho}(\pi^\rho - \varpi^\rho) \wedge \omega^k + \frac{1}{2}\sum (\gamma^i_{jk} - \Gamma^i_{jk})\omega^j \wedge \omega^k \tag{4}$$

and Cartan's lemma now implies

$$\sum a^i_{k\rho}(\pi^\rho - \varpi^\rho) + \frac{1}{2}\sum (\gamma^i_{kj} - \Gamma^i_{kj})\omega^j = \sum b^i_{kj}\omega^j,$$

where $b^i_{jk} = b^i_{kj}$.

Let the dimension of $G = r$, then the number of linearly independent elements among the $\{\sum a^i_{j\rho}\pi^\rho\}$ equals r, and hence there exist 1-forms

$$\sum a^{i_1}_{j_1\rho}\pi^\rho, \cdots, \sum a^{i_r}_{j_r\rho}\pi^\rho$$

which are linearly independent. As a consequence the matrix

$$(D^\sigma_\rho) = (a^{i_\sigma}_{j_\sigma\rho})$$

is invertible and can be applied to both sides of the equations

$$\sum a^{i_\sigma}_{j_\sigma \rho}(\pi^\rho - \varpi^\rho) = \sum (\Gamma^{i_\sigma}_{j_\sigma k} - \gamma^{i_\sigma}_{j_\sigma k} + b^{i_\sigma}_{j_\sigma k}) \wedge \omega^k.$$

This implies that $\pi^\rho = \varpi^\rho \bmod \text{base}$, or

$$\boxed{\pi^\rho - \varpi^\rho = \sum v^\rho_k \omega^k.} \tag{5}$$

Thus the indeterminancy is of the form

$$\begin{pmatrix} \varpi \\ \omega \end{pmatrix} = \begin{pmatrix} I & \nu \\ 0 & I \end{pmatrix} \begin{pmatrix} \pi \\ \omega \end{pmatrix},$$

and *any change of this type is Lie algebra compatible.* Now let us substitute the boxed equation back into (4) for the difference of the representations of the exterior derivatives and see what conditions are forced on the gammas. When we do this we find

$$0 = \sum a^i_{j\rho} v^\rho_k \, \omega^j \wedge \omega^k + \frac{1}{2} \sum (\gamma^i_{jk} - \Gamma^i_{jk}) \omega^j \wedge \omega^k.$$

In order to extract the conditions on the coefficents, it is necessary to skew-symmetrize them in (i, j), which yields

$$\boxed{\Gamma^i_{jk} = \gamma^i_{jk} + \sum (a^i_{j\rho} v^\rho_k - a^i_{k\rho} v^\rho_j).} \tag{6}$$

Now notice that the terms

$$-\sum (a^i_{j\rho} v^\rho_k - a^i_{k\rho} v^\rho_j)$$

are the image of the variables v^ρ_k under a linear operator **L** with constant coefficients $a^i_{j\rho}$. *This means that the torsion is well defined in a tensor space modulo the image of the operator* **L**. We will identify the structure of the domain and range shortly.

Let us now turn to a simple example of the use of Lie algebra compatible absorption.

Example 3. *Web geometry* (*continued*). We saw that this problem was equivalent to a determined equivalence problem which on the base spaces looked like

$$\Phi^* \begin{pmatrix} \Omega^1_V \\ \Omega^2_V \end{pmatrix} = \begin{pmatrix} u & 0 \\ 0 & u \end{pmatrix} \begin{pmatrix} \omega^1_U \\ \omega^2_U \end{pmatrix},$$

and which on the G-spaces, with G the group of 2×2 scalar matrices, looked like

$$\Phi^{1*} \begin{pmatrix} \Omega^1 \\ \Omega^2 \end{pmatrix} = \begin{pmatrix} \omega^1 \\ \omega^2 \end{pmatrix}.$$

Now, as we have already stated, we will perform the calculations only over $U \times G$. We start with an arbitrary representation of the the exterior derivatives,

$$d\begin{pmatrix}\omega^1\\ \omega^2\end{pmatrix} = \begin{pmatrix}\alpha & \beta\\ \gamma & \delta\end{pmatrix} \wedge \begin{pmatrix}\omega^1\\ \omega^2\end{pmatrix}.$$

Here we have written the matrix we called (Δ^i_j) in terms of 1-forms $\alpha, \beta, \gamma, \delta$. Since the Maurer–Cartan matrix for the two-dimensional scalar group has the form

$$dSS^{-1} = \begin{pmatrix}\varpi & 0\\ 0 & \varpi\end{pmatrix},$$

we immediately see three relations and hence have determined that

$$\alpha - \delta, \quad \beta, \quad \gamma$$

are principal components. The torsion terms created by the principal components are semibasic forms of top degree and hence are multiples of $\omega^1 \wedge \omega^2$. This means there are functions a and b defined on $U \times G$ which satisfy

$$d\begin{pmatrix}\omega^1\\ \omega^2\end{pmatrix} = \begin{pmatrix}\alpha & 0\\ 0 & \alpha\end{pmatrix} \wedge \begin{pmatrix}\omega^1\\ \omega^2\end{pmatrix} + \begin{pmatrix}a\\ b\end{pmatrix} \omega^1 \wedge \omega^2.$$

Now we want to modify α by multiples of ω^1 and ω^2 in such a way that we eliminate as much of the torsion as possible. Since

$$d\begin{pmatrix}\omega^1\\ \omega^2\end{pmatrix} = \begin{pmatrix}\alpha - a\omega^2 + b\omega^1 & 0\\ 0 & \alpha - a\omega^2 + b\omega^1\end{pmatrix} \wedge \begin{pmatrix}\omega^1\\ \omega^2\end{pmatrix},$$

this is a case in which all the torsion can be absorbed. *Although it is traditional to let the same symbol represent the modified forms,* let us temporarily break with that tradition for clarity. Thus let

$$\varphi = \alpha - a\omega^1 + b\omega^2.$$

The exterior derivative equations now have the form

$$d\begin{pmatrix}\omega^1\\ \omega^2\end{pmatrix} = \begin{pmatrix}\varphi & 0\\ 0 & \varphi\end{pmatrix} \wedge \begin{pmatrix}\omega^1\\ \omega^2\end{pmatrix},$$

and these equations uniquely determine the 1-form φ. The 1-forms ω^1, ω^2, and φ are now an invariant coframe on $U \times G$.

Exercise. Verify that φ is uniquely determined.

Let us see what we can deduce about the exterior derivative of φ. Since

$$\begin{aligned}0 = d^2\begin{pmatrix}\omega^1\\ \omega^2\end{pmatrix} &= \begin{pmatrix}d\varphi & 0\\ 0 & d\varphi\end{pmatrix} \wedge \begin{pmatrix}\omega^1\\ \omega^2\end{pmatrix} - \begin{pmatrix}\varphi & 0\\ 0 & \varphi\end{pmatrix} \wedge \begin{pmatrix}d\omega^1\\ d\omega^2\end{pmatrix}\\ &= \begin{pmatrix}d\varphi & 0\\ 0 & d\varphi\end{pmatrix} \wedge \begin{pmatrix}\omega^1\\ \omega^2\end{pmatrix} - \begin{pmatrix}\varphi & 0\\ 0 & \varphi\end{pmatrix} \wedge \begin{pmatrix}\varphi & 0\\ 0 & \varphi\end{pmatrix} \wedge \begin{pmatrix}\omega^1\\ \omega^2\end{pmatrix} = \begin{pmatrix}d\varphi \wedge \omega^1\\ d\varphi \wedge \omega^2\end{pmatrix},\end{aligned}$$

and $d\varphi$ is a 2-form, the above equations imply that there is a function c defined on $U \times G$ satisfying

$$d\varphi = c\, \omega^1 \wedge \omega^2.$$

In exactly the same manner, a function C can be defined on $V \times G$. These new functions have the property that the lift Φ^1 of any equivalence satisfies

$$C \circ \Phi^1 = c.$$

We will call such functions invariants.

Exercise. Verify this last equation.

Note that this whole analysis has been performed without knowing the explicit formula for any of the forms involved. This is what will be called an *abstract computation.* Once the abstract construction of invariants has been completed, there are two possibilities for the next move. First, it may be possible to prove theorems based on the structure equations

$$\begin{aligned} d\omega^1 &= \varphi \wedge \omega^1 \\ d\omega^2 &= \varphi \wedge \omega^2 \\ d\varphi &= c\, \omega^1 \wedge \omega^2. \end{aligned}$$

Second, it is usually desirable to have an explicit formula for the invariants in terms of the data of the problem and hence, if it is not too difficult, these need to be found.

Let me give an example of a theorem based on the abstract computation.

THEOREM. *The invariant c vanishes if and only if*

$$\frac{dy}{dx} = f(x,y) \ \textit{ is equivalent to } \ \frac{dY}{dX} = 1.$$

Proof. If $c = 0$, then $d\varphi = 0$ and by the Poincaré lemma (Sophomore version) we have

$$\varphi = dg,$$

where g is a function on $U \times G$. In addition since φ is a nonzero Maurer–Cartan form mod base, we know $g_u \neq 0$. Thus if we set

$$g(x,y,u) = \text{constant},$$

this determines a submanifold on which $\varphi = 0$ and satisfies the induced structure equations

$$d\omega^1 = 0, \qquad d\omega^2 = 0. \tag{7}$$

Explicitly,

$$\omega^1 = uF dx, \qquad \omega^2 = u dy,$$

hence (7) implies

$$uF = X'(x), \qquad u = Y'(y)$$

or

$$\begin{aligned} \omega^1 &= X'(x)dx = dX(x) \\ \omega^2 &= Y'(y)dy = dY(y). \end{aligned}$$

Now defining an admissible transformation by $h(x,y) = (X(x), Y(y))$, we have

$$h^*(dY - dX) = \omega^2 - \omega^1 = \omega^3 = u(x,y)(dy - f dx),$$

which means

$$\frac{dy}{dx} = f(x,y) \text{ is equivalent to } \frac{dY}{dX} = 1$$

as desired.

The explicit computations of formulas for invariants in terms of the data of the problem are called the *parametric calculations*. This part of the problem can be obtained from the abstract solution by use of a symbolic manipulation program like MAPLE, SMP, MACSYMA, MATHEMATICA, REDUCE, or SCRATCHPAD. In our case that would be pointless, since the calculations are so easy. The parametric representation of the lifted forms is

$$\omega^1 = uf dx, \qquad \omega^2 = u dy.$$

The exterior derivatives are easily calculated:

$$\begin{aligned} d\omega^1 &= \frac{du}{u} \wedge \omega^1 + uf_y dy \wedge dx = \left(\frac{du}{u} + \frac{f_y}{uf}\,\omega^2\right) \wedge \omega^1, \\ &= \left(\frac{du}{u} + \frac{(\ln|f|)_y}{u}\,\omega^2\right) \wedge \omega^1, \end{aligned}$$

and

$$d\omega^2 = \frac{du}{u} \wedge \omega^2 = \left(\frac{du}{u} + \frac{(\ln|f|)_y}{u}\,\omega^2\right) \wedge \omega^2.$$

Thus we see

$$\varphi = \frac{du}{u} + \frac{(\ln|f|)_y}{u}\,\omega^2.$$

In order to compute the invariant c we need the exterior derivative of φ. This is easiest if we note

$$\varphi = du/u + (\ln|f|)_y dy,$$

because then

$$d\varphi = (\ln|f|)_{yx} dx \wedge dy = \frac{1}{u^2 f}\,(\ln|f|)_{xy}\,\omega^1 \wedge \omega^2,$$

and we have the explicit parametric formula

$$c = \frac{1}{u^2 f}\,(\ln|f|)_{xy}.$$

Of course something other than personal satisfaction is usually gained when the explicit formulas for invariants are known. In this case, we can determine the explicit equivalence class of the equations with $c = 0$.

Since

$$c = 0 \quad \text{if and only if} \quad (\ln|f|)_{xy} = 0,$$

we integrate this wave equation to find $f(x, y) = e^{h(x)+g(y)}$, and see that the equivalence class consists of all equations of the form

$$\frac{dy}{dx} = e^{h(x)} e^{g(y)}.$$

We will have more to say about the case when $c \neq 0$ in Lecture 6.

Now let us return to the general procedure and recast the idea behind Lie algebra compatible absorption. *The goal is the solving of as many equations of the form*

$$\sum \pi^i_j \wedge \omega^j = \frac{1}{2} \sum \gamma^i_{jk} \omega^j \wedge \omega^k$$

as possible subject to the defining equations of the Lie algebra,

$$\sum b^i_{j\mu} \pi^j_i = 0, \qquad r + 1 \leq \mu \leq n^2.$$

Let us immediately look at an example from this viewpoint. In many equivalence problems, one of the generalized orthogonal groups *SO(p,q)* appear. These are the groups defined as the set of all $(p + q)$-dimensional matrices S which satisfy

$${}^tSQS = Q, \quad \det S = 1,$$

where Q is the nonsingular matrix with p ones and q minus ones on the diagonal. Let us find the defining relations on the matrix of Maurer–Cartan forms. This is done by taking the defining relations of the group G, and differentiating them. Thus

$${}^tdSQS + {}^tSQdS = 0,$$

or

$${}^tS^{-1}\,{}^tdSQ + QdSS^{-1} = 0,$$

or

$${}^t(dSS^{-1})Q + Q(dSS^{-1}) = 0.$$

Matters being so, if $\Pi = (\pi^k_j)$, the defining relations for the Lie algebra of *SO(p,q)* are

$${}^t\Pi\, Q + Q\, \Pi = 0.$$

There is a general algebraic result called the *fundamental theorem of pseudo-Riemannian geometry*, or in our context the *$SO(p,q)$ absorption theorem*, that answers the question of Lie algebra compatible absorption for this group.

This will be proved with a sequence of results each of which is a useful weapon for your algebraic arsenal.

LEMMA (S_3-*lemma*). *A three-index tensor C^i_{jk} that is symmetric in one pair of indices and skew-symmetric in another pair of indices is identically zero.*

(The usual proof is by an uninspired sequence of six identities. The following proof, which is the reason for the name of the lemma, was shown to me by J. Wolf.)

Proof. If C^i_{jk} is not identically zero, let l, m, and n be fixed indices with $C^l_{mn} \neq 0$. Let S_3 be the symmetric group on three letters and define a homomorphism

$$0 \longrightarrow K \longrightarrow S_3 \xrightarrow{\text{sign}} Z_2 \longrightarrow 0$$

with kernel K and image $Z_2 = \{\pm 1\}$, where for $\sigma \in S_3$

$$C^{\sigma(l)}_{\sigma(m)\sigma(n)} = \text{sign}(\sigma) C^l_{mn}.$$

The order of K is 3 and there is a unique subgroup of S_3 or order 3, the even permutations. The permutation

$$\sigma(l,m,n) = (\sigma(l), \sigma(m), \sigma(n)) = (n,l,m)$$

is even, but

$$C^{\sigma(l)}_{\sigma(m)\sigma(n)} = C^n_{lm} = -C^l_{nm} = -C^l_{mn}$$

shows that $\text{sign}(\sigma) = -1$, which contradicts $\sigma \in K$.

LEMMA (S_3-CARTAN LEMMA). *Let ω be an n-vector of linearly independent 1-forms, then the exterior system*

$$\xi \wedge \omega = 0, \qquad {}^t\xi + \xi = 0,$$

for ξ an $n \times n$ matrix of 1-forms, has $\xi = 0$ as the unique solution.

Proof. Applying Cartan's lemma to the first set of equations and directly using the second set of equations implies that the coefficients of $\xi = \sum A^i_{jk}\omega^k$ satisfy the S_3-lemma, and hence are zero.

THEOREM ($SO(p,q)$ ABSORPTION THEOREM). *Let $\omega = {}^t(\omega^1, \cdots, \omega^n)$ be linearly independent 1-forms, and let Q be a nonsingular symmetric matrix, then given*

$$\psi = \frac{1}{2}\sum \gamma^i_{jk}\omega^k, \text{ where } \gamma^i_{jk} = -\gamma^i_{kj},$$

there is a unique solution Π to the system of exterior equations

$$\Pi \wedge \omega = \psi \wedge \omega, \qquad {}^t\Pi\, Q + Q\,\Pi = 0.$$

Proof. Let $\xi = Q\Pi$ and let $\eta = Q\psi$, then the exterior system becomes

$$\xi \wedge \omega = \eta \wedge \omega, \qquad {}^t\xi + \xi = 0.$$

This is a system of $\frac{1}{2}n^2(n-1)$ inhomogeneous equations for $\frac{1}{2}n^2(n-1)$ unknowns and by the S_3-Cartan lemma, the homogeneous system has $\xi = 0$ as the unique solution. As a result, the inhomogeneous system also has a unique solution. To find this solution explicitly let

$$\eta = \frac{1}{2}\sum C^i_{jk}\omega^k,$$

then

$$\sum(\underbrace{\xi^i_k - \tfrac{1}{2}C^i_{kl}\omega^l + \overbrace{\tfrac{1}{2}C^k_{il}\omega^l}_{\text{skew}} + \underbrace{\tfrac{1}{2}C^l_{ik}\omega^l}_{\text{skew}}}^{\text{symmetric}}) \wedge \omega^k = 0$$

since the final two terms are symmetric in (k, l) and sum out to zero. The whole expression is now skew and the S_3-Cartan lemma gives

$$\xi^i_k = \sum \frac{1}{2}C^i_{kl}\omega^l - \frac{1}{2}C^k_{il}\omega^l - \frac{1}{2}C^l_{ik}\omega^l,$$

which satisfies the above system and proves existence.

As an application of these ideas, let us return to Example 1.

Example 1. *Riemannian geometry* (*continued*). On $U \times G$ where $G = O(n, \mathbf{R})$, we write

$$d\omega = \Delta \wedge \omega.$$

Since $SO(n, \mathbf{R}) = SO(n, 0)$, we have $Q = I$ and the relations on the Maurer–Cartan form matrix become,

$$dSS^{-1} + {}^t(dSS^{-1}) = 0.$$

As such $\Delta + {}^t\Delta$ are the principal components.

Now define $\delta = \frac{1}{2}(\Delta - {}^t\Delta)$ and $\psi = \frac{1}{2}(\Delta + {}^t\Delta)$ so that $\Delta = \delta + \psi$. Now this gives

$$d\omega = \delta \wedge \omega + \psi \wedge \omega,$$

and we want to absorb the torsion terms $\psi \wedge \omega$ into $\delta \wedge \omega$ while preserving the Lie algebra structure of δ.

This means we want to solve as many of the exterior equations

$$\Pi \wedge \omega = \psi \wedge \omega$$

as possible, subject to the Lie algebra conditions,

$${}^t\Pi + \Pi = 0.$$

This is just the special case of the $SO(p,q)$ absorption lemma, where $Q = I$. As such there is a unique solution Π to these exterior equations.

Now define

$$\varphi = \delta + \Pi$$

to produce the structure equations

$$d\omega = \varphi \wedge \omega, \quad \text{where } {}^t\varphi = -\varphi. \tag{8}$$

Since the construction of φ was unique, the 1-forms ω and φ are a G-invariant coframe on $U \times G$. The Lie algebra valued matrix of 1-forms Φ on $U \times G$ is, of course, the Levi–Civita connection form. Further differentiation of (8) immediately leads to the Riemann–Christoffel curvature matrix. We will return to this example in Lecture 6.

LECTURE 4

Reduction of the Structure Group

We have now seen two examples in which Lie algebra compatible absorption could be used to eliminate all the torsion terms. Unfortunately, this is not usually the case. To understand the next step in the algorithm we need to look further at the process of Lie algebra compatible absorption.

Let us introduce an n-dimensional vector space V with a basis $\{e_i\}$, which we will use to identify with $\mathbf{R}^n$ thought of as column vectors, and its dual space V^* with the dual basis $\{f^j\}$, which we will use to identify with $\mathbf{R}^n$ thought of as row vectors. We also need the basis $\{\varepsilon_\alpha\}$ for $T_e(G)$, which is dual to the basis $\{\pi^\rho\}$ of $T_e^*(G)$. Since we will be interested in the natural actions of G on these spaces, we will view $T_e(G)$ as the Lie algebra of G and write

$$T_e(G) \simeq \mathcal{G} \subset \operatorname{Hom}(V, V),$$

where under these identifications,

$$\varepsilon_\rho = \sum a^j_{i\rho} e_j \otimes f^i.$$

Matters being so, we view the linear map discussed in terms of components in the middle of Lecture 3 in the following way. Define

$$\mathcal{G} \otimes V^* \xrightarrow{\mathbf{L}} V \otimes \Lambda^2 V^*$$

by

$$\mathbf{L}\left(\sum \nu^\rho_k \varepsilon_\rho \otimes f^k\right) = -\frac{1}{2}\sum (a^i_{j\rho}\nu^\rho_k - a^i_{k\rho}\nu^\rho_j) e_i \otimes f^j \wedge f^k.$$

Associated to any linear map is its kernel-cokernel exact sequence

$$0 \longrightarrow \mathcal{G}^{(1)} \longrightarrow \mathcal{G} \otimes V^* \xrightarrow{\mathbf{L}} V \otimes \Lambda^2 V^* \longrightarrow \Pi_{\mathcal{G}} \longrightarrow 0.$$

Here of course

$$\mathcal{G}^{(1)} = \ker \mathbf{L} \text{ and } \Pi_{\mathcal{G}} = V \otimes \Lambda^2 V^* / \operatorname{im} \mathbf{L}.$$

The space $\mathcal{G}^{(1)}$ is called the first Lie algebra prolongation, and the space $\Pi_{\mathcal{G}}$ is subscripted since it depends on the Lie algebra structure of G and the representation $\rho : G \longrightarrow \mathrm{Aut}(V)$. The space $\Pi_{\mathcal{G}}$ is often denoted by $h^0(\mathcal{G})$ or $h^{0,2}(\mathcal{G})$ since it can be viewed as a Spencer cohomology construction.

Now let us define a map

$$g : U \times G \longrightarrow V \otimes \Lambda^2 V^*$$

by

$$g(u,S) = \sum \gamma^i_{jk}(u,S)\; e_i \otimes f^j \wedge f^k.$$

If we follow this map by the natural projection

$$V \otimes \Lambda^2 V^* \longrightarrow \Pi_{\mathcal{G}} \longrightarrow 0,$$

we obtain a mapping

$$\tau_U : U \times G \longrightarrow \Pi_{\mathcal{G}}.$$

This mapping τ_U is called the *intrinsic torsion.*

Now if

$$d\Omega^i = \sum a^i_{j\rho}\Pi^\rho \wedge \Omega^j + \frac{1}{2}\sum \Gamma^i_{jk}\Omega^j \wedge \Omega^k \quad \text{on } V \times G,$$

and

$$d\omega^i = \sum a^i_{j\rho}\pi^\rho \wedge \omega^j + \frac{1}{2}\sum \gamma^i_{jk}\omega^j \wedge \omega^k \quad \text{on } U \times G,$$

and there were an equivalence

$$\Phi^1 : U \times G \longrightarrow V \times G,$$

then the pullback equations from $V \times G$ become

$$d\omega^i = \sum a^i_{j\rho}\Phi^{1*}\Pi^\rho \wedge \omega^j + \frac{1}{2}\sum \Gamma^i_{jk} \circ \Phi^1 \omega^j \wedge \omega^k.$$

Using the argument leading to (5) in Lecture 3, this implies

$$\Phi^{1*}\Pi^\rho = \pi^\rho + \sum \nu^\rho_k \omega^k,$$

and

$$\frac{1}{2}\sum \Gamma^i_{jk} \circ \Phi^1 \omega^j \wedge \omega^k \equiv \frac{1}{2}\sum \gamma^i_{jk}\omega^j \wedge \omega^k \text{ mod im}\,\mathbf{L}.$$

As a result the diagram

$$\begin{array}{ccc} U \times G & \xrightarrow{\Phi^1} & V \times G \\ \tau_U \big\downarrow & & \big\downarrow \tau_V \\ \Pi_{\mathcal{G}} & \xrightarrow{\mathrm{Id}} & \Pi_{\mathcal{G}} \end{array}$$

commutes and the structure map is preserved under equivalences.

To simplify the discussion, let us suppress the components by defining the vector-valued forms

$$\varpi = \{\varpi^\rho\},\ \pi = \{\pi^\rho\},\ \nu = \{\nu^\rho_k\}$$

$$\Gamma = \{\Gamma^i_{jk}\},\ \gamma = \{\gamma^i_{jk}\}.$$

Thus in this notation the most general Lie algebra compatible absorption of (3) is defined by

$$\varpi = \pi + \nu\omega,$$

and this induces

$$\Gamma = \gamma - \mathbf{L}(\nu).$$

Now we try to solve as many of the equations

$$\mathbf{L}(\nu) = \gamma$$

as possible. Here and in the future we will identify the linear map **L** with the matrix of **L**. If the rank of the linear map **L** is s, then there exist s linearly independent rows of the corresponding linear equations, and we solve the system with the right-hand side of these s linear equations arbitrarily prescribed. Let us index such a set as follows:

$$\sum a^{i_1}_{j_1\rho}\nu^\rho_{k_1} - a^{i_1}_{k_1\rho}\nu^\rho_{j_1} = \gamma^{i_1}_{j_1k_1}$$

$$\vdots$$

$$\sum a^{i_s}_{j_s\rho}\nu^\rho_{k_s} - a^{i_s}_{k_s\rho}\nu^\rho_{j_s} = \gamma^{i_s}_{j_sk_s}.$$

If we fix such a set of components, let us consider the remaining ambiguity in our coframes. Using these choices, we define a vector space projection

$$\mathcal{P} : V \otimes \Lambda^2 V^* \longrightarrow \operatorname{im} L$$

by

$$\mathcal{P}\left(\sum h^i_{jk} e_i \otimes f^j \wedge f^k\right) = \sum_{\sigma=1}^{s} h^{i_\sigma}_{j_\sigma k_\sigma} e_{i_\sigma} \otimes f^{j_\sigma} \wedge f^{k_\sigma}.$$

Now given $H \in V \otimes \Lambda^2 V^*$, we abbreviate this map by

$$\mathcal{P}(H) = H_0,$$

and use this projection to decompose

$$H = H_0 + H_1$$

and write the corresponding decomposition $\gamma = \gamma_0 + \gamma_1$. By the previous remarks there is a ν' such that $\mathbf{L}(\nu') = \gamma_0$. If we restrict to Lie algebra compatible absorptions

$$\varpi = \pi + \kappa\omega$$

preserving the normalization, i.e., $\mathbf{L}(\kappa) = \gamma_0$, then

$$\kappa \in \nu' + \ker \mathbf{L}$$

or $\kappa = \nu' + \nu$ with $\nu \in \ker \mathbf{L}$. As such,

$$\begin{aligned}\varpi &= \pi + (\nu' + \nu)\omega \\ &= \pi + \nu'\omega + \nu\omega\end{aligned}$$

and making the initial Lie algebra compatible absorption,

$$\pi' = \pi + \nu'\omega,$$

we have the remaining possible normalized Lie algebra compatible absorption in the form $\varpi = \pi' + \nu\omega$ with

$$\boxed{\nu \in \ker \mathbf{L}.}$$

As a result the induced changes on the torsion coefficients become

$$\Gamma = \gamma - \mathbf{L}(\nu) = \gamma,$$

which means that the torsion coefficients are now well defined.

This last sequence of arguments was to show that the torsion coefficients can be made well defined. This is not the only way to do this since there are many ways to solve s of the equations $\mathbf{L}(\nu) = \gamma$ and hence to define the projection P. *In practice we want to choose P as G-equivariant as possible.* This difficulty does not arise in simple problems, but cannot be ignored, as we shall see in the examples of the Lagrangian field theories, and conformal geometry.

The *first structure equations* are the equations

$$\boxed{d\omega^i = \sum a^i_{jp}\pi^p \wedge \omega^j + \frac{1}{2}\sum \gamma^i_{jk}\omega^j \wedge \omega^k,}$$

where the coefficients γ^i_{jk} have been made unique by Lie algebra compatible absorption, and as a result the *ambiguity* of π is given by

$$(9) \qquad \begin{pmatrix} \varpi \\ \omega \end{pmatrix} = \begin{pmatrix} I & \nu \\ 0 & I \end{pmatrix} \begin{pmatrix} \pi \\ \omega \end{pmatrix}, \quad \text{where } \nu \in \ker \mathbf{L}.$$

Note that the first structure equations depend on the choice of projection P.

A natural algebraic question arises:

$$\text{“when is } \mathcal{G} \otimes V^* \xrightarrow{L} V \otimes \Lambda^2 V^* \text{ surjective?”}$$

because then all the torsion can be absorbed. This is a property of the Lie subalgebra $\mathcal{G} \subset \mathrm{Hom}(V, V)$ and was solved in the irreducible case by H. Weyl [W. 1922] and É. Cartan [C. 1922] and considered in the reducible case by Kobayashi and Nagano [K–N. 1965]. Robert Bryant pointed out that the paper [K–N. 1965] is not quite correct and supplied the proper statement and proof which can be found in Addendum 1 at the end of this lecture.

Now we need to build the group action into the analysis. In Lecture 2 we discussed the inclusion $T_e(G) \subseteq M(n, \mathbf{R})$, where we viewed $G \subseteq Gl(n, \mathbf{R})$ as the identity inclusion, i.e., G equals its image in $Gl(n, \mathbf{R})$. We generalize this viewpoint to fit representation theory and let

$$\rho : G \longrightarrow \mathrm{Aut}(V) \quad V \simeq \mathbf{R}^n$$

be the representation defining $G \subset Gl(n, \mathbf{R})$. This induces

$$\rho_* : T_e G \longrightarrow \mathrm{Hom}(V, V)$$

and results in

$$\rho_* : \mathcal{G} \longrightarrow \mathrm{Hom}(V, V).$$

Associated to ρ is the dual or contragredient representation

$$\rho^\dagger : G \longrightarrow \mathrm{Aut}(V^*)$$

characterized by

$$\langle \rho(S)v, \rho^\dagger(S)f \rangle = \langle v, f \rangle \text{ for } v \in V, f \in V^*,$$

and given explicitly in terms of matrices

$$\rho^\dagger(S) = {}^t\rho(S)^{-1}.$$

If we let $\rho(S) = (s^i_j)$ and $\rho(S)^{-1} = (\sigma^i_j)$, then the natural G-action on $V \otimes \Lambda^2 V^*$ is given by

$$\rho \otimes \wedge^2 \rho^\dagger : G \longrightarrow \mathrm{Aut}(V \otimes \Lambda^2 V^*).$$

This action is explicitly given on the components h^i_{jk} of a tensor by

$$\rho \otimes \wedge^2 \rho^\dagger(S) h^i_{jk} = \sum \sigma^p_j \sigma^q_k h^m_{pq} s^i_m.$$

There is also the group action on itself by inner automorphisms

$$\mathrm{Int} : G \longrightarrow Diff(G), \text{ where } \mathrm{Int}(S)T = STS^{-1}.$$

This induces a group representation called the *adjoint representation*

$$\mathrm{Ad} : G \longrightarrow \mathrm{Aut}\,(\mathcal{G}), \text{ where } \mathrm{Ad}\,(S) = (\mathrm{Int}\,(S))_*.$$

If we recall

$$\mathcal{G} = T_e(G) \subset M(n, \mathbf{R}),$$

and let $[\alpha(t)]$ be an equivalence class of curves defining a tangent vector in T_eG, then

$$\mathrm{Ad}(S)[\alpha(t)] = \mathrm{Int}_*(S)([\alpha(t)]) = [S\alpha(t)S^{-1}] = S[\alpha(t)]S^{-1}.$$

Dually there is the *coadjoint* representation

$$\mathrm{Ad}^\dagger : G \longrightarrow \mathrm{Aut}\,(T_e^*(G)),$$

a representation we will need in Lecture 10.

There is also a natural G-action on $\mathcal{G} \otimes V^*$, given by

$$\mathrm{Ad} \otimes \rho^\dagger : G \longrightarrow \mathrm{Aut}\,(\mathcal{G} \otimes V^*),$$

and a natural G-action

$$\rho_* \circ \mathrm{Ad} \otimes \rho^\dagger : G \longrightarrow \mathrm{Aut}\,(\mathrm{Hom}\,(V, V) \otimes V).$$

If $\{\epsilon_\sigma\}$ is the basis of $T_eG \simeq \mathcal{G}$ dual to $\{\pi^\rho\}$ then

$$\rho_*\epsilon_\sigma = a^j_{i\sigma} \in \mathrm{Hom}(V, V).$$

If we let $\rho(s) = \sigma^i_m$, $\rho^t(s) = s^k_j$, and

$$\mathrm{Ad}_\rho(s) \circ \rho_*\epsilon_\sigma = \sum \alpha^\sigma_\tau\, a^i_{j\sigma}$$

then $\rho_* \circ \mathrm{Ad}\, s(\epsilon_\sigma) = \rho(s)\rho_*(\epsilon_\sigma)\rho(s)^{-1}$, because

$$\sum \alpha^\sigma_\tau\, a^i_{j\sigma} = \sum s^k_j a^l_{k\tau} \sigma^i_l$$

and as a result

$$\begin{aligned}
\rho \otimes \wedge^2 \rho^*(s) \circ \mathbf{L}(\nu) &= \sum \sigma^i_m \left(\sum a^m_{p\tau}\nu^\tau_q - a^m_{q\tau}\nu^\tau_p \right) s^p_j s^q_k \\
&= \sum \alpha^\sigma_\tau a^i_{j\sigma} \nu^\tau_q s^q_k - \alpha^\sigma_\tau a^i_{k\sigma} \nu^\tau_p s^p_j \\
&= a^i_{j\sigma}(\alpha^\sigma_\tau \nu^\tau_q s^q_k) - a^i_{k\sigma}(\alpha^\sigma_\tau \nu^\tau_p s^p_j) \\
&= \mathbf{L}(\rho_* \circ \ \mathrm{Ad} \ \otimes \rho^\dagger(s)\nu),
\end{aligned}$$

which proves that the linear map $\mathbf{L}$ is actually a mapping of G-modules

$$\mathrm{Hom}\,(V, V) \otimes V^* \xrightarrow{\mathbf{L}} V \otimes \Lambda^2 V^*.$$

As a result the sequence

$$0 \longrightarrow \mathcal{G}^{(1)} \longrightarrow \mathcal{G} \otimes V^* \xrightarrow{L} V \otimes \Lambda^2 V^* \longrightarrow \Pi_{\mathcal{G}} \longrightarrow 0$$

is an exact sequence of G-modules.

Since the quotient projection is a G-module mapping, we see

$$\tau_U(u, CS) = (\rho \otimes \wedge^2 \rho^\dagger)(C)\ \tau_U(u, S)$$

and hence deduce that the image of a fiber of $U \times G$ over U is an orbit of the action of G on $\Pi_{\mathcal{G}}$.

Now we come to an important definition.

An equivalence problem is of *first-order constant type* if $\tau_U(U \times G)$ is a single orbit of G on $\Pi_{\mathcal{G}}$.

If an equivalence problem is not of first-order constant type, then there are two possibilities. First, there could be a finite number of orbits, in which case the method can be applied one orbit at a time. Second, it may happen that the orbits are not finite in number. Cartan actually indicates how this case might be handled, but the added complexity in exposition and subtlety prohibits the consideration of this difficulty in any generality.

Now we assume that the equivalence problem is of first-order constant type. If we choose a fixed vector, *usually a particular normal form in the image of the structure map*, say

$$\tau_0 = \tau_U(u_0, S_0)\ \in\ \tau_U(U \times G),$$

then there will be an isotropy group of this vector,

$$G_{\tau_0} = \{C \in G \mid \rho \otimes \Lambda^2 \rho^\dagger(C)\tau_0 = \tau_0\}.$$

Since $\tau_U(U \times G)$ is an orbit, and there is only one orbit, it follows that $\tau_U(U \times G)$ is the orbit of τ_0. Now since G acts transitively on $\tau_U(U \times G)$ it can be identified with the homogeneous space G/G_{τ_0}. The structure map

$$\tau_U : U \times G \longrightarrow G/G_{\tau_0}$$

has constant rank, hence $\tau_u^{-1}(\tau_0)$ is a manifold. The point $\tau_U(u, e)$ is on the orbit, hence for each u there is a $C(u) \in G$ such that

$$(\rho \otimes \wedge^2 \rho^\dagger)(C(u))\tau_U(u, e) = \tau_0.$$

Therefore

$$\tau_U^{-1}(\tau_0) = \{(u, C(u)\, G_{\tau_0}) | u \in U\}$$

is a manifold which submerses onto U. A τ_0-modified coframe is a section of $\tau_U^{-1}(\tau_0)$. The Implicit Function Theorem implies that there is a local section, $\Gamma(u) = (u, \beta_U(u))$, where

$$\beta_U : U \longrightarrow G,$$

which satisfies

$$\tau_U(u, \beta_U(u)) = \tau_U(\Gamma(u)) = \tau_0.$$

Utilizing this map $\beta_U : U \longrightarrow G$ we can construct a τ_0-*modified coframe* by taking the coframe $\beta_U \omega_U$.

Note if there is an equivalence Φ^1 with $\Phi^1(u_0, S_0) = (v_0, T_0)$ then

$$\tau_0 = \tau_U(u_0, S_0) = \tau_V \circ \Phi^1(u_0, S_0) = \tau_V(v_0, T_0).$$

With this preparation we can now state the most important and historically worst-treated result in the subject.

THEOREM (REDUCTION OF THE STRUCTURE GROUP). *A mapping*

$$\Phi : U \longrightarrow V$$

induces a G-equivalence if and only if Φ *induces a* G_{τ_0}*-equivalence between* τ_0*-modified coframes given by*

$$\beta_U \omega_U \quad \textit{and} \quad \beta_V \Omega_V.$$

Proof. Given a solution Φ of the original equivalence problem, we have

$$\Phi^*(\beta_V \Omega_V) = (\beta_V \circ \Phi)\gamma_{VU}\omega_U = (\beta_V \circ \Phi)\gamma_{VU}\beta_U^{-1}\beta_U\omega_U$$

and if we define $\alpha_{VU} = (\beta_V \circ f)\gamma_{VU}\beta_U^{-1}$, we must show $\alpha_{VU} \in G_{\tau_0}$.

Let the lifted equivalence of Φ be Φ^1, then we have

$$\begin{aligned}
\tau_0 &= \tau_U(u, \beta_U(u)) = \tau_V \circ \Phi^1(u, \beta_U(u)) = \tau_V(\Phi(u), \beta_U(u)\gamma_{VU}^{-1}) \\
&= \tau_V(\Phi(u), \beta_U(u)\gamma_{VU}^{-1}(\beta_V \circ \Phi(u))^{-1}(\beta_V \circ \Phi(u))) \\
&= (\rho \otimes \Lambda^2\rho^\dagger)(\beta_U\gamma_{VU}^{-1}(\beta_V \circ \Phi(u))^{-1})\tau_V(\Phi(u), \beta_V \circ \Phi(u)) \\
&= (\rho \otimes \Lambda^2\rho^\dagger)(\beta_U\gamma_{VU}^{-1}(\beta_V \circ \Phi)^{-1})\tau_0,
\end{aligned}$$

hence $\beta_U\gamma_{VU}(\beta_V \circ \Phi)^{-1}$ is contained in G_{τ_0}, but then the inverse $(\beta_V \circ \Phi)\gamma_{VU}\beta_U^{-1}$ is contained in G_{τ_0}, as desired.

Conversely, given a solution for the equivalence problem

$$\Phi^*(\beta_V\Omega_V) = \alpha_{VU}(\beta_U\omega_U) \quad \text{with} \quad \alpha_{VU} \in G_{\tau_0},$$

then

$$\Phi^*\Omega_V = (\beta_V \circ \Phi)^{-1}\alpha_{VU}\beta_U\omega_U.$$

Since $(\beta_V \circ \Phi)^{-1}$ and β_V are in G, α_{VU} is in G_{τ_0}, and G_{τ_0} is a subgroup of G

$$\gamma_{VU} = (\beta_V \circ \Phi)^{-1}\alpha_{UV}\beta_U \in G$$

as required.

This is great theoretically, since the smaller the group the easier the problem is likely to be.

There is an important class of problems, which was pointed out by Robert Bryant, that are not of the first-order constant type, but are still amenable to group reductions. This will be discussed in Addendum 2 at the end of the lecture.

A practical point we now need to address is the method to choose τ_0. Since working on a quotient space is annoying, it is natural to ask if we can study the G-action on $\Pi_{\mathcal{G}}$ by lifting calculations to $V \otimes \Lambda^2 V^*$.

As such, let us choose a vector-space splitting

$$0 \longrightarrow \Pi_{\mathcal{G}} \xrightarrow{\sigma} V \otimes \Lambda^2 V^*.$$

If we let $H \in V \otimes \Lambda^2 V^*$ and [H] its equivalence class in $\Pi_{\mathcal{G}}$, we may define a vector-space splitting by

$$\sigma([H]) = H_1,$$

where H_1 was the complement to the projection P on $\operatorname{Im} \mathbf{L}$ defined in the first part of this lecture. This is equivalent to making the choice of P defined earlier.

The splitting described above cannot usually be chosen to be a G-module splitting, but should be chosen to be a G-module splitting if possible. Natural combinations of the equations $\mathbf{L}(\nu) = \gamma$, such as contractions, are preferred, if they exist, since then the associated splitting will be a G-module splitting.

The secret to choosing $\tau_0 \in \Pi_{\mathcal{G}}$ in a canonical way is to find a normal form for the orbit $\tau_U(U \times G)$ in $\Pi_{\mathcal{G}}$. The classical approach, that is, before the 1980s and excluding Cartan, was to work on the group level. This required actually parameterizing the group G or knowing deep special properties about G. The most efficient approach is to work at the Lie algebra level.

The group action on the structure tensor can be uncovered by computing the relations

$$d^2\omega^i = 0$$

modulo subspaces in the base, and using linear combinations if necessary to solve for the torsion terms

$$d\gamma^i_{jk} \quad \text{mod base.}$$

This is the information needed to deduce the G-action on the structure tensor and to choose an appropriate normalization. In his theory of Répère Mobile, Cartan called such a normalization *the first-order normalization.*

It is a fact that the more representation theory you know the easier it is to solve complicated equivalence problems. If you know how to decompose

$$\Pi_{\mathcal{G}} = \bigoplus_{\alpha} W_\alpha$$

into irreducibles then you can focus on relatively simple actions.

Although this representation theory may seem hard to apply, *often irreducible subspaces can be spotted from* $d\gamma^i_{jk}$ mod base. *Irreducible components of the structure tensor are often produced by introducing block matrix decompositions of G or by looking at the structure equations modulo known invariant subspaces.*

Infinitesimal group actions.

A representation $\rho : G \to \operatorname{Aut} V$ induces a linear map

$$\rho_* : T_e(G) \to T_I \operatorname{Aut} V,$$

where I is the identity transformation on V, defined on an equivalence class of curves $[\alpha(t)]$ with $\alpha(0) = e$ defining a tangent vector by the formula

$$\rho_*[\alpha(t)] = [\rho(\alpha(t))].$$

Using the identifications

$$T_e G \subset M(n, \mathbf{R}) \quad \text{and} \quad T_I \operatorname{Aut} V \simeq \operatorname{Hom}(V, V)$$

this mapping is defined by

$$\rho_* \left(\frac{d\alpha}{dt} \right) = \frac{d\rho(\alpha(t))}{dt}.$$

Since this holds for all curves α passing through the identity we have

$$\rho_*(dS|_e) = d\rho(S)|_e.$$

At first sight this formula seems unnatural, but if we recall

$$dS \in M(n, \mathbf{R}) \otimes T^*_S(G)$$

we see that the action ρ_* above is really $\rho_* \otimes \mathbf{1}$.

Now let $v_0 \in V$ and consider the vector valued function on G given by

$$\nu(S) = \rho(S) v_0.$$

Then

$$\begin{aligned} d\nu &= d(\rho(S) v_0) = d\rho(S) \rho(S)^{-1} \rho(S) \nu_0 \\ &= \rho_* \, (dS S^{-1}) \, \nu \end{aligned}$$

hence

$$\boxed{d\nu - \rho_* \, (dS S^{-1}) \, \nu = 0.}$$

Thus the infinitesimal group action is expressed in the boxed equation, and standard Lie theory tells us this is sufficient to reconstruct the local group action as long as G is simply connected (see [Wa. 1971]). Let us look at the standard Lie algebra actions.

The contragredient representation. Here $\rho^\dagger(S) = {}^tS^{-1}$ and

$$\rho^\dagger_*\left(\frac{d\alpha(t)}{dt}\right)v_0 = \left(\frac{d\rho^\dagger(\alpha(t))}{dt}\right)v_0 = \left(\frac{d\,{}^t\alpha(t)^{-1}}{dt}\right)v_0$$
$$= -{}^t\left(\alpha(t)^{-1}\frac{d\alpha(t)}{dt}\alpha(t)^{-1}\right)v_0$$

which at $t = 0$ gives

$$= -\frac{{}^td\alpha}{dt}(0)v_0.$$

The similarity representation.

$$\rho(S)v_0 = Sv_0S^{-1}$$
$$\rho_*\left(\frac{d\alpha(t)}{dt}\right)v_0 = \frac{d}{dt}(\alpha(t)v_0\alpha(t)^{-1})$$
$$= \frac{d\alpha(t)}{dt}v_0\alpha(t)^{-1} + \alpha(t)v_0\frac{d}{dt}(\alpha(t)^{-1})$$

which at $t = 0$ gives

$$= \frac{d\alpha}{dt}(0)v_0 - v_0\frac{d\alpha}{dt}(0).$$

The tensor product of two representations. Given $\phi : G \to \operatorname{Aut} V$ and $\psi : G \to \operatorname{Aut} W$, then $\phi \otimes \psi : G \to \operatorname{Aut}(V \otimes W)$ is defined by

$$(\phi \otimes \psi)(S) = \phi(S) \otimes \psi(S).$$

Thus we look at the action

$$(\phi \otimes \psi)_*\left(\frac{d\alpha(t)}{dt}\right) = \frac{d}{dt}(\phi(\alpha(t)) \otimes \psi(\alpha(t)))$$
$$= \frac{d}{dt}(\phi(\alpha(t))) \otimes \psi(\alpha(t)) + \phi(\alpha(t)) \otimes \frac{d}{dt}(\psi(\alpha(t)))$$

which at $t = 0$ gives

$$= \phi_*\left(\frac{d\alpha}{dt}\right) \otimes 1 + 1 \otimes \psi_*\left(\frac{d\alpha}{dt}\right).$$

Translational action. This is not a vector representation but arises as part of a vector representation, e.g.,

$$\begin{pmatrix} I & x \\ 0 & 1 \end{pmatrix} \begin{pmatrix} v_0 \\ 1 \end{pmatrix} = \begin{pmatrix} v_0 + x \\ 1 \end{pmatrix}.$$

Hence let

$$v(x) = f(x)v_0 = v_0 + x$$

then

$$f_* \left(\frac{d\alpha(t)}{dt} \right) v_0 = \frac{df(\alpha(t))}{dt} v_0 = \frac{d}{dt}(v_0 + \alpha(t))$$

which at $t = 0$ gives

$$= \frac{d\alpha}{dt}.$$

Determinantal action. This is the representation

$$\rho(S) = (\det S)^m.$$

Then

$$\rho_* \left(\frac{d\alpha}{dt} \right) = \frac{d}{dt}(\det \alpha(t))^m$$
$$= m\,(\det \alpha(t))^{m-1} \frac{d \det \alpha(t)}{dt}.$$

Using $d \log(\det S) = \operatorname{tr}(dSS^{-1})$ at $S = e$ we have

$$\rho_* \left(\frac{d\alpha}{dt} \right) = m \operatorname{tr} \left(\frac{d\alpha}{dt} \right).$$

Example. To see these ideas in an applications, let us assume a block decomposition of the structure equations in the following form

$$d \begin{pmatrix} \omega \\ \theta \\ \eta \end{pmatrix} = \begin{pmatrix} \pi_1 & 0 & 0 \\ \beta_1 & 0 & 0 \\ \beta & \beta_2 & \pi_2 \end{pmatrix} \wedge \begin{pmatrix} \omega \\ \theta \\ \eta \end{pmatrix} + \begin{pmatrix} b\,\eta \wedge \omega \\ A\,\eta \wedge \omega \\ 0 \end{pmatrix},$$

where A is a matrix and b is a vector, then

$$d^2 \begin{pmatrix} \omega \\ \theta \\ \eta \end{pmatrix} = 0$$

leads to

$$dA - A\pi_2 + \pi_1 A \equiv 0 \quad \operatorname{mod}(\omega, \theta, \eta)$$
$$db + \pi_2 b - A\,\beta_1 \equiv 0 \quad \operatorname{mod}(\omega, \theta, \eta).$$

Hence, since there are no relations on π_1, π_2 and β_1, we deduce the action of G on A is by

$$PAQ^{-1},$$

and the action of G on b is by left multiplication by Q and by a translation, but the nature of the translation depends on the nature of A. If A is invertible then it is an arbitrary translation in the subspace in which b lies. In this case we could normalize $A = I$ and $b = 0$.

There is one additional Lie algebra representation we will need later; this is the representation induced by the coadjoint action.

Coadjoint action. The coadjoint representation

$$\mathrm{Ad}^\dagger : G \longrightarrow \mathrm{Aut}(T_e^*(G), T_e^*(G))$$

induces a Lie algebra map

$$\mathrm{ad}^\dagger : \mathcal{G} \longrightarrow \mathrm{Hom}(T_e^*(G), T_e^*(G)).$$

This action is important in identifying the Lie group that preserves the structure equations produced at the end of the method. This will be taken up in the last lecture.

Let $X = [\alpha(t)]$ be a right invariant vector field, hence $\alpha(t)$ is a 1-parameter group of left translations which we denote by $\exp tX$. Then given ϖ a right invariant Maurer–Cartan form,

$$\begin{aligned}\langle \mathrm{ad}^\dagger X, \varpi \rangle &= \langle \mathrm{Ad}_*^\dagger[\alpha(t)], \varpi \rangle = \frac{d}{dt} L_{\alpha(t)}^* \circ R_{\alpha(t)}^* \varpi \\ &= \frac{d}{dt} L_{\alpha(t)}^* \varpi = \frac{d}{dt}(\exp\, tX)^* \varpi = \mathcal{L}_X \varpi,\end{aligned}$$

where $\mathcal{L}_X$ is the Lie derivative. Using the Cartan formula, relating the Lie derivative, exterior derivative and interior product we see

$$\langle \mathrm{ad}^\dagger X, \varpi \rangle = X \lrcorner d\varpi + d(X \lrcorner \varpi) = X \lrcorner d\varpi,$$

since $X \lrcorner \varpi = \text{constant}$.

Now if we choose $\{X_\rho\} \in \mathcal{G}$ dual to a basis $\{\varpi^\rho\}$ of right invariant Maurer–Cartan forms and use the structure equations of G, we see

$$X_\rho \lrcorner d\varpi^\sigma = X_\rho \lrcorner \frac{1}{2} \sum C_{\alpha\beta}^\sigma \varpi^\alpha \wedge \varpi^\beta = \sum C_{\rho\beta}^\sigma \varpi^\beta,$$

and therefore,

$$\mathrm{ad}^\dagger X_\rho = (C_{\rho\beta}^\sigma) \in M(n, \mathbf{R}).$$

Addendum 1. The question

$$\text{“when is } \mathcal{G} \otimes V^* \xrightarrow{\mathrm{L}} V \otimes \wedge^2 V^* \text{ surjective?”}$$

has the following solution.

Case I. If $\mathcal{G}$ acts irreducibly on V then either $\mathcal{G}$ contains a sub-algebra h which is the sub-algebra preserving a nondegenerate symmetric bilinear form $\langle\ ,\ \rangle$ on V or else dim V=2 and there exists a constant $\lambda > 0$ so that, in a basis, we have

$$\left\{ \begin{pmatrix} x & -\lambda x \\ \lambda x & x \end{pmatrix} \middle| \, x \in \mathbf{R} \right\},$$

(note that, as λ varies over $(0, \infty)$, these algebras are all inequivalent).

Case II. If $\mathcal{G}$ acts reducibly on V, then $\mathcal{G}$ preserves a subspace of dimension 1 (exactly) and, choosing bases appropriately, the possibilities are (for $\dim V = n + r$):

(1)

$$\mathcal{G} = \left\{ \begin{pmatrix} a & b \\ 0 & A \end{pmatrix} \middle| \, a \in \mathbf{R}, b \in {}^t(\mathbf{R}^n), A \in gl(n, \mathbf{R}) \right\}$$

(2)

$$\mathcal{G}_\lambda = \left\{ \begin{pmatrix} \lambda(\mathrm{tr}A) & b \\ 0 & A \end{pmatrix} \middle| \, A \in gl(n, \mathbf{R}), b \in {}^t(\mathbf{R}^n) \right\}, \quad \lambda \in \mathbf{R}$$

(3)

$$\mathcal{G} = \left\{ \begin{pmatrix} a & 0 \\ 0 & b \end{pmatrix} \middle| \, a, b \in \mathbf{R} \right\}$$

if $\dim V = 2$

(4)

$$\mathcal{G}_\lambda = \left\{ \begin{pmatrix} a & 0 \\ 0 & \lambda a \end{pmatrix} \middle| \, a \in \mathbf{R} \right\}$$

$\dim V = 2,\ \lambda \neq 0.$

In Case II, it is easy to see that the four cases exhaust the possibilities as follows: If $W \subseteq V$ is fixed by $\mathcal{G}$ and $\dim W = p > 0$ and $\dim V = p + q > p$, then we may choose bases so that

$$W = (x^1, \cdots, x^p, 0, \cdots, 0).$$

If we had torsion equations of the form

$$1 \leqslant i \leqslant p, \qquad d\omega^i = \sum \frac{1}{2}\gamma^i_{jk}\, \omega^j \wedge \omega^k + \gamma^i_{j\alpha}\, \omega^j \wedge \omega^\alpha + \frac{1}{2}\gamma^i_{\alpha\beta}\, \omega^\alpha \wedge \omega^\beta$$

$$p < \alpha \leqslant p + q, \qquad d\omega^\alpha = \sum \frac{1}{2}\gamma^\alpha_{jk}\, \omega^j \wedge \omega^k + \gamma^\alpha_{j\beta}\, \omega^j \wedge \omega^\beta + \frac{1}{2}\gamma^\alpha_{\beta\gamma}\, \omega^\beta \wedge \omega^\gamma$$

then $p > 1$ implies that the γ^α_{jk} cannot be absorbed since $\mathcal{G}$ is a sub-algebra of

$$\begin{pmatrix} * & * \\ 0 & * \end{pmatrix} = gl(V, W).$$

If $p = 1$, then $\gamma^\alpha_{jk} \equiv 0$ of course, so we need to worry about $\gamma^\alpha_{j\beta}$ and $\gamma^\alpha_{\beta\gamma}$. But $\gamma^\alpha_{1\beta}$ can be absorbed for arbitrary $\gamma^\alpha_{1\beta}$ if and only if the lower right $q \times q$ block is everything. Thus $\mathcal{G}$ must contain the algebra

$$\begin{pmatrix} 0 & 0 \\ 0 & gl(\mathcal{G}) \end{pmatrix}.$$

This will suffice to absorb all of the $d\omega^\alpha$ torsion. Now, if $q > 1$, then there will be nontrivial $\gamma^i_{\alpha\beta}$ terms. Thus, $\mathcal{G}$ cannot be in

$$\begin{pmatrix} * & 0 \\ 0 & gl(\mathcal{G}) \end{pmatrix}$$

Since

$$\begin{pmatrix} 0 & 0 \\ 0 & gl(\mathcal{G}) \end{pmatrix}$$

acts irreducibly on

$$\begin{pmatrix} 0 & * \\ 0 & 0 \end{pmatrix},$$

it follows that $\mathcal{G}$ contains

$$\begin{pmatrix} 0 & * \\ 0 & gl(\mathcal{G}) \end{pmatrix}.$$

This algebra is enough to absorb all the torsion, and the subcases (i) and (ii) are the only algebras between this one and $gl(V, W)$. The case $q = 1$ is exceptional, for then $\gamma^i_{\alpha\beta}$ vanishes identically and we may then take $b = 0$ which leads to subcases (iii) and (iv).

Addendum 2. In spite of Cartan's indications on how to proceed, the general case of equivalence problems resulting in an infinite number of orbits may not even be approachable. There is, however, a situation, which occurs naturally, where group reduction is applicable.

Let $G \subseteq GL(n, \mathbf{R})$ be a Lie group, then a subgroup $H \subseteq G$ is said to be of *stabilizer type* if there exists an element $v \in \Pi_{\mathcal{G}}$ so that $H = \{g \in G | gv = v\}$. Note that any subgroup conjugate to one of stabilizer type is also of stabilizer type since, if $H = \{g \in G | gv = v\}$, then $aHa^{-1} = \{g \in G | g(av) = av\}$. If $H \subseteq G$ is of stabilizer type, we may let $\mathrm{W}_H = \{v \in \Pi_{\mathcal{G}} | Hv = v\}$. Then W_H is clearly a nontrivial vector subspace of $\Pi_{\mathcal{G}}$. We let $\mathring{W}_H = \{v \in W_H | gv = v \Rightarrow g \in H\}$. Then $\mathring{W}_H$ is easily seen to be an open dense subset of W_H.

Note that if $v \in \mathring{W}_H$ and $gv \in \mathring{W}_H$ then $gHg^{-1} = H$, so $g \in N(H)$ (equal to the normalizer of H in G). Thus

$$(Gv) \cap \mathring{W}_H = N(H)v.$$

It easily follows that if we define a map $\mu : G \times \mathring{W}_H \longrightarrow \Pi_{\mathcal{G}}$ by

$$\mu(g, v) = gv,$$

then the fibers of this map are the orbits of $N(H)$ under the action $(g, v)\,n = (gn, n^{-1}v)$ where $n \in N(H)$. If we let $\Sigma_H = \mu(G \times \mathring{W}_H)$, then Σ_H consists exactly of those elements of $\Pi_{\mathcal{G}}$ whose stabilizers are conjugate to H. Thus, Σ_H is a smooth submanifold diffeomorphic to $(G \times \mathring{W}_H)/N(H)$. In fact, we have a fibration

$$\begin{array}{ccc} N(H) & \longrightarrow & G \times \mathring{W}_H \\ & & \downarrow{\scriptstyle \mu} \\ & & \Sigma_H. \end{array}$$

A submanifold $S \subseteq \mathring{W}_H$ will be said to be a *section* if, for all $v \in S$, $Gv \cap S = v$ and S is transverse to Gv in $\mathring{W}_H$ at v (it is not obvious that sections exist). Note that, by definition, $\dim S = \dim \mathring{W} H - \dim G + \dim H$. We will say that S is *maximal* if it is not a proper subset of any other section in $\mathring{W}_H$. We will say that S is *affine* if S is an open subset of an affine subspace and *linear* if S is an open subset of a linear subspace of $\mathring{W}_H$. One method of constructing an affine section is to take $v \in \mathring{W}_H$ and a linear subspace $\mathcal{L} \subseteq \mathring{W}_H$ so that $\mathcal{L}$ is a complement to $T_v(N(H)v)$. Then, if the $N(H)$-orbits are closed in $\mathring{W}_H$, an open subset of $v + \mathcal{L}$ will form a section.

Given a triple (G, H, S) as above, we will say that a G-structure $B \subseteq \mathcal{F}(M^n)$ *admits an H-reduction of type S* if $\tau_U : B \longrightarrow \Pi_{\mathcal{G}}$ has image in $G \cdot S$, where τ_U is the intrinsic torsion function. In this case, we may define an H-structure $B' \subseteq B$ by letting $B' = \tau_U^{-1}(S)$ (by hypothesis, τ_U is transverse to S). This is the most general case in which a smooth reduction to an H-structure can be accomplished. Note that a necessary condition for this (but not sufficient) is that $\tau_U(B)$ have its image in Σ_H.

LECTURE 5

The Inductive Step

The contents of the first four lectures can be visually summed up in the flowchart shown in Fig. 1.

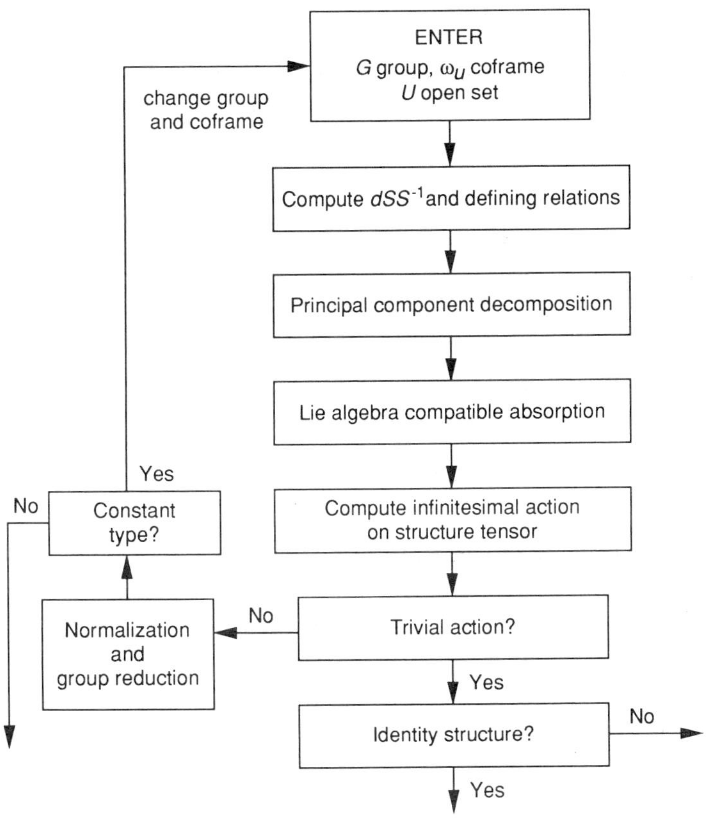

FIG. 1. *Flowchart. Loop* A.

The inductive step is simply to keep going down Loop A with the new group G_{τ_0} and the new coframe. Let us begin with an example since the last lecture was all theory.

Example 4. *One-dimensional time-fixed Newton's equations* (*continued*).

$$\begin{pmatrix} \omega^1 \\ \omega^2 \\ \omega^3 \end{pmatrix} = \begin{pmatrix} 1 & 0 & 0 \\ 0 & 1 & 0 \\ -z & z & w \end{pmatrix} \begin{pmatrix} \omega_U^1 \\ \omega_U^2 \\ \omega_U^3 \end{pmatrix}.$$

If $d\omega = \Delta \wedge \omega$ then since

$$d\begin{pmatrix} 1 & 0 & 0 \\ 0 & 1 & 0 \\ -z & z & w \end{pmatrix} \begin{pmatrix} 1 & 0 & 0 \\ 0 & 1 & 0 \\ z/w & -z/w & 1/w \end{pmatrix} = \begin{pmatrix} 0 & 0 & 0 \\ 0 & 0 & 0 \\ -* & * & dw/w \end{pmatrix},$$

where $* = dz - z/w\, dw$, the principal components are given by

$$\Delta_{11}, \Delta_{12}, \Delta_{13}, \Delta_{21}, \Delta_{22}, \Delta_{23}, \text{ and } \Delta_{31} + \Delta_{32}.$$

This could be seen without explicit parametrization by noting

$$(w, z) \to \begin{pmatrix} 1 & 0 & 0 \\ 0 & 1 & 0 \\ -z & z & w \end{pmatrix}$$

gives a faithful representation of the semi-direct product of $\mathbf{R}^*$ and $\mathbf{R}$ with connecting homomorphism given by right multiplication, i.e.,

$$(w, z)(w', z') = (ww', z + wz').$$

The general structure of Lie algebras of semidirect products then implies

$$\tilde{\omega}(S) = \begin{pmatrix} 0 & 0 & 0 \\ 0 & 0 & 0 \\ -* & * & ** \end{pmatrix}.$$

Often the nature of the problem leads to integrability conditions that force the principal components to be zero and not just congruent to zero mod base. In this case the special information consists of

$$d\omega^1 = d\omega_U^1 = d(dx) = 0,$$

so here $\Delta_{11}, \Delta_{12}, \Delta_{13}$ can be taken to be zero; and

$$d\omega^2 \wedge \omega^2 = d\omega_U^2 \wedge \omega_U^2 = -\frac{dy'}{y'} \wedge \omega^2 \wedge \omega^2 = 0,$$

hence $d\omega^2$ is a multiple of ω^2 and we can take Δ_{21} and Δ_{23} to be zero. Therefore we may assume

$$d\begin{pmatrix} \omega^1 \\ \omega^2 \\ \omega^3 \end{pmatrix} = \begin{pmatrix} 0 & 0 & 0 \\ 0 & \alpha & 0 \\ \gamma & \beta & \delta \end{pmatrix} \wedge \begin{pmatrix} \omega^1 \\ \omega^2 \\ \omega^3 \end{pmatrix}$$

with principal components of order 1

$$\alpha, \ \beta + \gamma.$$

Since

$$\alpha = a\,\omega^1 + b\,\omega^2 + c\,\omega^3 \quad \text{and} \quad \gamma = -\beta + e\,\omega^1 + f\,\omega^2 + g\,\omega^3$$

we can rewrite this equation in the form

$$d\begin{pmatrix}\omega^1\\ \omega^2\\ \omega^3\end{pmatrix} = \begin{pmatrix}0 & 0 & 0\\ 0 & 0 & 0\\ -\beta & \beta & \delta\end{pmatrix} \wedge \begin{pmatrix}\omega^1\\ \omega^2\\ \omega^3\end{pmatrix} + \begin{pmatrix}0\\ a\,\omega^1\wedge\omega^2 + b\,\omega^3\wedge\omega^2\\ e\,\omega^1\wedge\omega^2 + g\,\omega^3\wedge\omega^2\end{pmatrix}.$$

The terms $e\,\omega^1\wedge\omega^2$ and $g\,\omega^3\wedge\omega^2$ can be absorbed by replacing δ by $\delta - g\,\omega^2$ and β by $\beta - e\,\omega^2$. The remaining terms $a\,\omega^1\wedge\omega^2$ and $b\,\omega^3\wedge\omega^2$ are well defined.

We will not introduce a new name for these modified forms β and δ. As usual we compute

$$\begin{aligned} 0 &= d^2\omega^2\\ &= da\wedge\omega^1\wedge\omega^2 - a\,\omega^1\wedge(c\,\omega^3\wedge\omega^2) + dc\wedge\omega^3\omega^2 + c\,d\omega^3\wedge\omega^2\\ &\qquad - c\,\omega^3\wedge(a\,\omega^1\wedge\omega^2). \end{aligned}$$

If we wedge with ω^3 we get

$$da\wedge\omega^1\wedge\omega^2\wedge\omega^3 + c\,\beta\wedge\omega^1\wedge\omega^2\wedge\omega^3 = 0$$

and if we wedge with ω^1 we get

$$db\wedge\omega^3\wedge\omega^2\wedge\omega^1 + c\,\delta\wedge\omega^3\wedge\omega^2\wedge\omega^1 = 0$$

which implies

$$\left.\begin{aligned} da + c\,\beta &\equiv 0\\ dc + c\,\delta &\equiv 0 \end{aligned}\right. \quad \text{mod base.}$$

Exercise. Show $c \neq 0$. (Hint: look at $d\omega^2\wedge\omega^1$.)

Now we can scale c to -1 and then translate a to zero to give

$$d\omega^2 = \omega^2\wedge\omega^3$$

and

$$\beta, \delta \equiv 0 \quad \text{mod base.}$$

Thus we have found new principal components which are called principal components of order 2. The group $G^{(1)}$ is now reduced to $\{e\}$, and we have new torsion terms

$$d\omega^3 = l\,\omega^1\wedge\omega^2 + m\,\omega^2\wedge\omega^3 + n\,\omega^1\wedge\omega^3.$$

Not all of this new torsion is nontrivial because there is an integrability condition arising from the reduced equation for ω^2. This results in

$$\begin{aligned} 0 = d^2\omega^2 &= d\omega^2 \wedge \omega^3 - \omega^2 \wedge d\omega^3 \\ &= -\omega^2 \wedge d\omega^3 \\ &= -n\,\omega^2 \wedge \omega^1 \wedge \omega^3 \end{aligned}$$

which implies $n = 0$. Thus there are two basic invariants l and m and structure equations

$$\begin{aligned} d\omega^1 &= 0 \\ d\omega^2 &= \omega^2 \wedge \omega^3 \\ d\omega^3 &= l\,\omega^1 \wedge \omega^2 + m\,\omega^2 \wedge \omega^3. \end{aligned}$$

Exercise. Do the parametric calculation and show

$$l = -\frac{1}{y'}\frac{\partial f}{\partial x} \quad \text{and} \quad m = \frac{\partial f}{\partial y'} - \frac{2f}{y'}.$$

Note that we continued the procedure without having to do nearly as much work as the first pass through Loop A to the group reduction. *In particular, note that although we reduced the group to G_{τ_0}, where τ_0 was determined by $b = -1$ and $A = 0$, we never had to explicitly determine G_{τ_0} or find the explicit change of initial coframe. Further the new principal components were handed to us for free.* This may not seem very spectacular since we know $G_{\tau_0} = \{e\}$, but this feature would have been the same even if there were further normalizations and reductions. This makes the abstract solution of equivalence problems' magnitudes easier than the parametric solutions. As mentioned before, given the abstract solution it may be possible to find the parametric solution by using a symbolic manipulation program, but now the change of frame corresponding to the reduction must be explicitly considered.

Another advantage to the inductive nature of the procedure is that the problem need not be completed to get interesting information. *In fact, each time around Loop* A *produces invariants involving information at one higher derivative's complexity.* Thus, if the interpretation of invariants loses significance when they involve derivatives of a certain order $N+1$ and all invariants have not yet been determined, the procedure can be stopped after N times around Loop A. Let us consider an example where once around Loop A gives interesting information even though the full solution requires three times around Loop A.

Example 6. *Geometry of the integral* (*continued*). The group G was seen to be

$$G = \begin{pmatrix} A_1 & 0 & 0 \\ b_1 & D & 0 \\ B & b_2 & A_2 \end{pmatrix}$$

with $\det D = 1$. The Maurer–Cartan matrix for $S \in G$ is

$$dSS^{-1} = \begin{pmatrix} * & 0 & 0 \\ * & \lambda & 0 \\ * & * & * \end{pmatrix},$$

with $\operatorname{tr}\lambda = 0$. If

$$d\begin{pmatrix}\theta\\ \omega\\ \eta\end{pmatrix} = \begin{pmatrix}\pi_1 & \alpha_1 & \alpha_2\\ \beta_1 & \delta & \alpha_3\\ \beta & \beta_2 & \pi_2\end{pmatrix} \wedge \begin{pmatrix}\theta\\ \omega\\ \eta\end{pmatrix},$$

We see the principal components of order 1 are α_1, α_2, α_3, $\operatorname{tr}\delta$. The forms $d\theta$ and $d\omega$ have no terms quadratic in η, hence

$$\alpha_2,\ \alpha_3 \equiv 0 \mod (\theta,\ \omega).$$

As such we may write

$$d\begin{pmatrix}\theta\\ \omega\\ \eta\end{pmatrix} = \begin{pmatrix}\pi_1 & 0 & 0\\ \beta_1 & \bar{\delta} & 0\\ \beta & \beta_2 & \pi_2\end{pmatrix} \wedge \begin{pmatrix}\theta\\ \omega\\ \eta\end{pmatrix} + \begin{pmatrix}\alpha_1 \wedge \omega + \alpha_2 \wedge \eta\\ \frac{1}{m}\operatorname{tr}\delta\,\mathbf{I} \wedge \omega + \alpha_3 \wedge \eta\\ 0\end{pmatrix}$$

where $\bar{\delta} = \delta - \frac{1}{m}\operatorname{tr}\delta\,\mathbf{I}$, hence $\operatorname{tr}\bar{\delta} = 0$.

If we write

$$\alpha_1 = r\theta + {}^t\omega J + l\eta, \quad \alpha_2 = n\theta + {}^t\omega N, \quad \alpha_3 = p\theta + {}^t\omega L,$$
$$(1/m)\operatorname{tr}\delta = q\theta + {}^t\omega P + {}^t\eta V$$

and set

$$\bar{\pi}_1 = \pi_1 - r\omega - n\,\eta, \qquad \bar{\beta} = \beta - q\omega - p\eta$$

we see that we can absorb the terms with r, n, p, and q.

Since P only occurs as ${}^t\omega P\omega$ we may assume ${}^tP = -P$, and deduce $\operatorname{tr} P = 0$. Thus setting

$$\Delta = \bar{\delta} + {}^t\omega P, \quad {}^t\eta \bar{W}\omega = {}^t\eta V\omega + {}^t\omega L\eta, \quad {}^t\omega M\eta = l\eta \wedge \omega + {}^t\omega N\eta$$

which is an *example of a natural absorption*, we have

$$d\begin{pmatrix}\theta\\ \omega\\ \eta\end{pmatrix} = \begin{pmatrix}\pi_1 & 0 & 0\\ \beta_1 & \Delta & 0\\ \beta & \beta_2 & \pi_2\end{pmatrix} \wedge \begin{pmatrix}\theta\\ \omega\\ \eta\end{pmatrix} + \begin{pmatrix}{}^t\omega J\omega + {}^t\omega M\eta\\ {}^t\eta W\,\omega\\ 0\end{pmatrix} \tag{9}$$

and taking d^2 gives

$$\left.\begin{aligned} dJ - {}^t\Delta J - J\Delta + \pi_1 J - \frac{1}{2}(M\beta_2 - {}^t\beta_2{}^tM) &\equiv 0\\ dM - \pi_1 M - M\pi_2 - {}^t\Delta M &\equiv 0\\ dW + \beta_1 + {}^t\pi_2 W &\equiv 0\end{aligned}\right\} \quad \text{mod base.}$$

A parametric calculation shows

$$M = -\frac{1}{L^{1/m}} A_1\,{}^tD^{-1}A_2^{-1} \neq 0,$$

hence we can normalize $M = \mathbf{I}$, $J = 0$, and $W = 0$. The next rounds of the loop are quite complicated, and in fact have only been completed for $m = 1$ (see [B-G. 1990]) and $m = 2, n = 1$ (see [G-S. 1985]), and as such we will stop here after looking at the special case of $m = 1$, the particle Lagrangians.

In this case the torsion (9) before normalization can be written in the simplified form

$$d\begin{pmatrix}\theta\\ \omega\\ \eta\end{pmatrix} = \begin{pmatrix}\pi_1 & 0 & 0\\ \beta_1 & 0 & 0\\ \beta & \beta_2 & \pi_2\end{pmatrix} \wedge \begin{pmatrix}\theta\\ \omega\\ \eta\end{pmatrix} + \begin{pmatrix}A\\ b\\ 0\end{pmatrix} \eta\wedge\omega.$$

with the induced action

$$\left.\begin{aligned} dA &\equiv \pi_1 A - A\pi_2\\ db &\equiv \beta_1 - b\pi_2\end{aligned}\right\} \quad \text{mod base.}$$

A parametric calculation shows that A is a nonsingular multiple of $(\partial^2 l/\partial p_\alpha \partial p_\beta)$ and hence $\det A \neq 0$ is the regular problem in the calculus of variations. Let us look at the positive definite case. With this assumption we can normalize

$$A = \mathbf{I}, \qquad b = 0$$

and find

$$d\begin{pmatrix}\theta\\ \omega\\ \eta\end{pmatrix} = \begin{pmatrix}\pi_1 & 0 & 0\\ \beta_1 & 0 & 0\\ \beta & \beta_2 & \pi_2\end{pmatrix} \wedge \begin{pmatrix}\theta\\ \omega\\ \eta\end{pmatrix} + \begin{pmatrix}\eta\wedge\omega\\ 0\\ 0\end{pmatrix}$$

with new principal components

$$\pi_1 - \pi_2, \ \beta_1.$$

This means there is no group action left on ω and hence ω is invariant.

A parametric calculation gives

$$\omega = L\,dx + \sum \frac{\partial L}{\partial p_\alpha}(dz^\alpha - p_\alpha\,dx).$$

This is the *Cartan form* or the *Hilbert Invariant Integral*, which is the key tool needed to establish Hamilton's principal in the Calculus of Variations. It is quite typical that the method of equivalence will uncover subtle invariant forms and conservation laws without need for deep insight into the problem.

Overdetermined equivalence problems.

Let G, $G' \subset Gl(n, \mathbf{R})$ be two subgroups and let ω_U, θ_U and Ω_V, Θ_V be two pairs of coframes on U and V, respectively. We want to find necessary and sufficient conditions that there exists a diffeomorphism

$$\Phi : U \to V$$

such that

$$\Phi^* \Omega_V = \gamma_{VU} \omega_U, \qquad \Phi^* \Theta_V = \alpha_{VU} \theta_U,$$

where $\gamma_{VU} \in G$ and $\alpha_{VU} \in G'$.

As before, this problem can be lifted to finding

$$\Phi^1 : U \times G \times G' \to V \times G \times G',$$

where

$$\omega|_{(u,S,A)} = S\omega_U, \qquad \theta|_{(u,S,A)} = A\theta_U$$
$$\Omega|_{(v,T,B)} = T\Omega_V, \qquad \Theta|_{(v,T,B)} = B\Theta_V$$

satisfy

$$\Phi^{1*}\Omega = \omega, \qquad \Phi^{1*}\Theta = \theta.$$

Exercise. Verify this by mimicking the argument for the determined equivalence problem (see [G–S. 1986]).

The idea is to use the group actions on the lifts to $U \times G \times G'$ to normalize the problem and reduce to a subgroup $G_1 \subset G$ (or equivalently to an isomorphic subgroup $G_1' \subset G'$). Since we are dealing with coframes there are nonsingular matrices m_U and M_V such that

$$\theta_U = m_U \omega_U, \qquad \Theta_V = M_V \Omega_V.$$

Similarly there are nonsingular matrices $\mathbf{m}$ and M such that

$$\theta = \mathbf{m}\omega, \qquad \Theta = M\Omega,$$

where in fact

$$\mathbf{m} = A m_U S^{-1}, \qquad M = B M_V T^{-1}.$$

The lift Φ^1 is easily seen to satisfy

$$M \circ \Phi^1 = \mathbf{m},$$

so we have a *zeroth-order invariant.*

The reduction of structure group technique suggests that we use the $G \times G'$ action to normalize this invariant matrix of functions. In particular, the natural goal is to normalize some submatrix to constant values or, if possible, the whole matrix to constants. This is not always possible and leads to equivalence problems with *invariant functions.* Although Cartan even indicated how to deal with this twist (see Appendix), we will not, again because the extra complexity would cloud the general ideas.

If we normalize $\mathbf{m}$ to m_0 then we have to restrict to $S_0 \in G$ and $A_0 \in G'$, satisfying

$$A_0 m_U S_0^{-1} = m_0,$$

and this induces a reduction of the lift of ω to $S_0\omega_U$. Now choose a fixed coframe ω_U^0 at each point with $\omega_U^0 \in S_0\omega_U$ and define a subgroup of G by

$$G_1 = \{\sigma \in G \mid \sigma\omega_U^0 \in S_0\omega_U\}, \quad \text{where } A_0 m_U S_0^{-1} = m_0.$$

The reduction of the structure group technique now goes through (see [G–S. 1986] for details) and the original problem is equivalent to solving

$$\Phi^*\Omega_V^0 = \beta_{VU}\omega_U^0, \qquad \beta_{UV} \in G_1,$$

where Ω_V^0 has been chosen, component by component, exactly as ω_U^0 has been chosen.

Example 3. *Web geometry* (*continued*). We saw that $y' = f(x,y)$ under $\Phi(x,y) = (\phi(x), \psi(y))$ is an overdetermined equivalence problem with

$$\omega = \begin{pmatrix} u & 0 \\ 0 & v \end{pmatrix} \begin{pmatrix} dx \\ dy \end{pmatrix}, \qquad \theta = \begin{pmatrix} a & 0 \\ b & c \end{pmatrix} \begin{pmatrix} dy - f\,dx \\ dx \end{pmatrix},$$

thus

$$\theta = \begin{pmatrix} a & 0 \\ b & c \end{pmatrix} \begin{pmatrix} -f & 1 \\ 1 & 0 \end{pmatrix} \begin{pmatrix} 1/u & 0 \\ 0 & 1/v \end{pmatrix} \omega = \underbrace{\begin{pmatrix} \dfrac{-af}{u} & \dfrac{a}{v} \\ \dfrac{-bf+c}{u} & \dfrac{b}{v} \end{pmatrix}}_{\mathbf{m}} \omega.$$

Normalize

$$\mathbf{m} = \begin{pmatrix} \dfrac{-af}{u} & \dfrac{a}{v} \\ \dfrac{-bf+c}{u} & \dfrac{b}{v} \end{pmatrix}$$

by setting

$$v = a, \qquad u = c$$
$$0 = b, \qquad u = af$$

which gives

$$\mathbf{m} \to \begin{pmatrix} -1 & 1 \\ 1 & 0 \end{pmatrix}.$$

This induces

$$\tilde{\omega} = \begin{pmatrix} af & 0 \\ 0 & a \end{pmatrix} \begin{pmatrix} dx \\ dy \end{pmatrix}.$$

A fixed coframe in this family is given by $a = 1$ or

$$\omega_U^0 = \begin{pmatrix} f\,dx \\ dy \end{pmatrix}.$$

The elements of G which have $\tilde{\omega}$ as the ω_U^0 orbit satisfy

$$u\,f\,dx = a\,f\,dx, \qquad v\,dy = a\,dy,$$

hence $v = a$ and $u = a$ and thus $u = v$. This shows

$$\tilde{\omega} = \underbrace{\begin{pmatrix} u & 0 \\ 0 & u \end{pmatrix}}_{G_1'} \underbrace{\begin{pmatrix} f\,dx \\ dy \end{pmatrix}}_{\omega_U^0}.$$

which recaptures the reduction found by a special technique in Lecture 1.

Example 4. *Time-fixed Newton's equations* (*continued*). We saw $y'' = F(x, y, y')$ under $\Phi(x, y) = (x, \phi(x))$ is an overdetermined problem with

$$\omega = \begin{pmatrix} 1 & 0 & 0 \\ 0 & u & 0 \\ a & b & c \end{pmatrix} \begin{pmatrix} dx \\ dy \\ dy' \end{pmatrix}, \qquad \theta = \begin{pmatrix} v & 0 & 0 \\ z & w & 0 \\ l & m & n \end{pmatrix} \begin{pmatrix} dy - y'dx \\ dy' - f\,dx \\ dx \end{pmatrix}$$

then

$$\begin{aligned} \theta &= \begin{pmatrix} v & 0 & 0 \\ z & w & 0 \\ l & m & n \end{pmatrix} \begin{pmatrix} -y' & 1 & 0 \\ -f & 0 & 1 \\ 1 & 0 & 0 \end{pmatrix} \begin{pmatrix} 1 & 0 & 0 \\ 0 & 1/u & 0 \\ -a/c & -b/uc & 1/c \end{pmatrix} \omega \\ &= \underbrace{\begin{pmatrix} -y'v & v/u & 0 \\ -y'z - w(f + a/c) & (z - bw/c)/u & w/c \\ -y'l - m(f + a/c) + n & (l - bm/c)/u & m/c \end{pmatrix}}_{\mathbf{m}} \omega. \end{aligned}$$

Setting

$$v = 1/y', \quad v = u \ \text{ implies } \ u = 1/y',$$

and

$$c = w, \quad z = \frac{bw}{c} \ \text{ implies } \ b = z,$$

and

$$m = 0, \quad l = \frac{bm}{c} \ \text{ implies } \ l = 0.$$

The (3,1) position $-y'l - m(f + a/c) + n$ equals n with $n \neq 0$ and the (2,1) position $-y'z - w(f + a/c)$ equals $-y'z - wf - a$; therefore $\mathbf{m}$ has the form

$$\begin{pmatrix} -1 & 1 & 0 \\ * & 0 & 1 \\ n & 0 & 0 \end{pmatrix}.$$

Now since $n \neq 0$ set

$$n = 1 \quad \text{and} \quad a = -y'z - wf,$$

which normalizes

$$\mathbf{m} \to \begin{pmatrix} 1 & 1 & 0 \\ 0 & 0 & 1 \\ 1 & 0 & 0 \end{pmatrix}.$$

This induces

$$\begin{aligned} \tilde{\omega} &= \begin{pmatrix} 1 & 0 & 0 \\ 0 & 1/y' & 0 \\ -y'z - wf & z & w \end{pmatrix} \begin{pmatrix} dx \\ dy \\ dz \end{pmatrix} \qquad (\text{note } w \neq 0) \\ &= \begin{pmatrix} dx \\ dy/y' \\ z(dy - y'dx) + w(dy' - f\,dx) \end{pmatrix}. \end{aligned}$$

A fixed coframe in this family is given by

$$z = 0, \qquad w = 1$$

or

$$\omega_U^0 = \begin{pmatrix} dx \\ dy/y' \\ dy' - f\,dx \end{pmatrix}.$$

The elements of G which have $\tilde{\omega}$ as the ω_U^0 orbit have

$$u\,\frac{dy}{y'} = \frac{dy}{y'} \quad \text{which implies} \quad u = 1$$

and

$$a\,dx + b\,\frac{dy}{y'} + c(dy' - f\,dx) = z(dy - y'dx) + w(dy' - f\,dx)$$

or $c = w$, $a = -zy'$, $b = zy'$, which implies $a = -b$. Thus

$$\tilde{\omega} = \underbrace{\begin{pmatrix} 1 & 0 & 0 \\ 0 & 1 & 0 \\ -b & b & c \end{pmatrix}}_{G_1} \underbrace{\begin{pmatrix} dx \\ dy/y' \\ dy' - f\,dx \end{pmatrix}}_{\omega_U^0},$$

which again recaptures the reduction found by a special technique in Lecture 1.

LECTURE 6

e-Structures

After enough iterations of Loop A, we will come to the situation where the action of the reduction of the group on torsion is trivial. If G has been reduced to the identity then $\{\omega^1, \cdots, \omega^m\}$ no longer involve group parameters and define an invariant coframe on U. The Newton equation example was such a case. If G has been reduced to a group with $\mathcal{G}^{(1)} = 0$, then $\{\omega^1, \cdots, \omega^m, \pi^1, \cdots, \pi^r\}$ defines an invariant coframe on $U \times G$. This occurred for both Web geometry and Riemannian geometry.

In either case, we have an equivalence problem, one on U, the other on $U \times G$ with an invariant coframe, and hence equivalence problems with $G = e$. In modern language this is the problem of finding invariants of e-structures, where an *e-structure* is simply a space with a specified coframe.

Cartan called the equivalence problem with the trivial group the restricted equivalence problem. This has a complete solution which will now be presented (see [C. 1908], [S. 1964]).

Thus assume we are given two coframings (U, ω) and (V, Ω) and want to find necessary and sufficient conditions that

$$\Phi^* \Omega = \omega.$$

The structure equations have the form

$$d\omega^i = \sum \gamma^i_{jk} \omega^j \wedge \omega^k \quad \text{and} \quad d\Omega^i = \sum \Gamma^i_{jk} \Omega^j \wedge \Omega^k.$$

If there were an equivalence, then $\mathcal{G}^{(1)} = 0$ implies

$$\gamma^i_{jk} = \Gamma^i_{jk} \circ \Phi.$$

Given a coframing ω, there are natural "covariant derivatives" for a function $f : U \to \mathbf{R}$, defined by

$$df = \sum f_{|i} \omega^i.$$

Similarly, given Ω and a function $g : V \to \mathbf{R}$, we define

$$dg = \sum g_{;i} \Omega^i.$$

As such

$$\begin{aligned}\sum \gamma^i_{jk|l}\omega^l &= d\gamma^i_{jk} = d(\Gamma^i_{jk}\circ\Phi) = \Phi^*(d\Gamma^i_{jk})\\ &= \Phi^*\left(\sum \Gamma^i_{jk;l}\Omega^l\right) = \sum \Gamma^i_{jk;l}\circ\Phi\,\Phi^*\Omega^l\\ &= \sum \Gamma^i_{jk;l}\circ\Phi\,\omega^l\end{aligned}$$

and as a result

$$\gamma^i_{jk|l} = \Gamma^i_{jk;l}\circ\Phi.$$

This last argument is of course inductive for higher covariant derivatives of the structure functions.

Now define

$$F_s(\omega) = \{\gamma^i_{jk}, \gamma^i_{jk|i_1}, \cdots, \gamma^i_{jk|i_1\cdots|i_{s-1}};\ 1 \le i, j, k, i_1, \cdots, i_{s-1} \le n\},$$

which we view as a lexicographically ordered set. This is a set of invariants of the s-jet of the e-structure ω. (Note the definition of $F_s(\omega)$ differs from Sternberg [S. 1964] in that his index s is one less than the order of the jet.)

Two natural invariants of this set may be defined. Let

$$k_s = \operatorname{rank}\{dF_s(\omega)\},$$

where we mean the dimension of the span of the differentials which occur in the ordered set $F_s(\omega)$.

Thus k_s is an integer valued function on U. The *order* of the e-structure at $p \in U$ is the smallest j such that

$$k_j(p) = k_{j+1}(p).$$

If j is the order of the e-structure at p, then the *rank* of the e-structure at p is the dimension

$$\rho(p) = k_j(p).$$

Note $0 \le j \le n$, the case $j = 0$ occurring when the structure tensor has constant coefficients and $\rho = n$ occurring if and only if one invariant function is added at each jet level.

As such an e-structure is said to be *regular of rank* ρ *at* p if the rank ρ is constant in a neighborhood of p. In this case there exist functions $\{f_1, \cdots, f_\rho\}$ defined in a neighborhood of p where

$$f_1, \cdots, f_\rho \in F_j(\omega), \qquad df_1 \wedge \cdots \wedge df_\rho \ne 0,$$

and any function $g \in F_j(\omega)$ satisfies

$$dg \wedge df_1 \wedge \cdots \wedge df_\rho = 0.$$

Note that these $f_1, \cdots, f_\rho$ can be extended to a coordinate system about p. We denote such a coordinate system by $\{x^i\}$ where $x^i = f_i$ for $1 \le i \le \rho$.

Exercise. If an e-structure is regular of order j at p, there is a neighborhood of p such that

$$k_l = k_j, \qquad l > j.$$

THEOREM (EQUIVALENCE OF e-STRUCTURES). *Let ω and Ω be regular e-structures of the same order j and rank ρ. Let*

$$h_U : U \to \mathbf{R}^m \quad \text{and} \quad h_V : V \to \mathbf{R}^m$$

be extensions of an independent set of elements of $F_j(\omega)$ and $F_j(\Omega)$ to coordinate systems constructed from identical lexicographic choices of indices. Define

$$\sigma = h_V^{-1} \circ h_U : U \to V,$$

then necessary and sufficient conditions that there exist

$$\Phi : U \to V \quad \text{with} \quad \Phi^*\Omega = \omega$$

are that

$$F_{j+1}(\Omega) \circ \sigma = F_{j+1}(\omega)$$

as lexicographically ordered sets.

This theorem is nicely done in [S. 1964], the proof there and here is Cartan's, but our treatment is more leisurely. The difficulty is that we must produce the map Φ from the e-structures ω and Ω. Cartan had a method to produce maps which F. Warner calls the technique of the graph (see [Wa. 1971]). As the name implies, the idea is to solve for the graph of Φ

$$\Gamma(\Phi) : U \to U \times V$$

defined by

$$\Gamma(\Phi)(u) = (u, \Phi(u)).$$

We actually used a special case of this technique when we characterized immersions in groups up to right multiplications in Lecture 2.

PROPOSITION (TECHNIQUE OF THE GRAPH). *Let (U, ω), (V, Ω) be e-structures with U connected and let π_U, π_V be the natural projections of $U \times V$ on U and V. If the differential system*

$$\Delta_{U \times V} = \pi_U^*\omega - \pi_V^*\Omega$$

is completely integrable, then there is a map $\Phi : U \to V$ satisfying

$$\Phi^*\Omega = \omega;$$

further, if the condition $\Phi(p) = q$ is added, the map is unique.

Proof. Given the hypothesis and the condition $\Phi(p) = q$, we consider the unique leaf $\mathcal{L}$ through (p, q). We claim

$$\pi_{U*} : T(\mathcal{L}) \to T(U)$$

is an isomorphism. In order to see that π_{U*} is injective we assume $\pi_{U*}(X) = 0$, then, since $X \in T(\mathcal{L})$,

$$\begin{aligned} 0 = \Delta_{U\times V}(X) &= \langle \pi_U^* \omega, X\rangle - \langle \pi_V^* \Omega, X\rangle \\ &= -\langle \pi_V^* \Omega, X\rangle, \end{aligned}$$

but Ω is a basis, hence $\pi_{V*}X = 0$ and $X = 0$. Since $\dim T(\mathcal{L}) = n = \dim T(U)$ and π_{U*} is injective, it is an isomorphism. Now we apply the Inverse Function Theorem to deduce that $\pi_U|_{\mathcal{L}}$ is a *local* diffeomorphism for some open set $W \in \mathcal{L}$. Extending this to a global argument is a subtle business, but important when possible. See F. Warner's book [Wa. 1971] for some examples.

As such we may define

$$\Phi = \pi_V \circ \pi_U^{-1},$$

where the maps are defined as follows:

$$U \overset{\pi_U^{-1}}{\to} \mathcal{L} \overset{\pi_V}{\to} V.$$

We claim Φ has the desired property. Let $z \in T(U)$ then

$$\begin{aligned} 0 = \Delta_{U\times V}\left(\pi_{U*}^{-1} z\right) &- \langle \pi_U^* \omega, \pi_{U*}^{-1} z\rangle - \langle \pi_V^* \Omega, \pi_{U*}^{-1} z\rangle \\ &= \langle \omega, z\rangle - \langle \Phi^* \Omega, z\rangle \end{aligned}$$

and since z was arbitrary

$$\Phi^* \Omega = \omega.$$

If we were given a second solution

$$g : U \to V \quad \text{with } g^* \Omega = \omega,\ g(p) = q$$

then the graph of g

$$\Gamma(g)(u) = (u, g(u)) = (I \times g)(u)$$

satisfies

$$\begin{aligned} \Gamma(g)^* \Delta_{U\times V} &= (\pi_U \circ \Gamma(g))^* \omega - (\pi_V \circ \Gamma(g))^* \Omega \\ &= I^* \omega - g^* \Omega \\ &= 0. \end{aligned}$$

In particular, then, the graph of g agrees with the leaf $\mathcal{L}$, and the uniqueness of the Frobenius theorem implies

$$\Gamma(g) = \Gamma(\Phi)$$

or

$$\Phi = g.$$

Note that the general solution of the problem without the initial data $\Phi(p) = q$ depends on the n parameters corresponding to the specification of the coordinates of q. A diffeomorphism Φ satisfying

$$\Phi^* \omega = \omega$$

is called an *automorphism of the e-structure.*

COROLLARY. *Let U be a connected open set with an e-structure ω; then any automorphism with a fixed point is the identity.*

Proof. Any automorphism Φ is uniquely determined by initial conditions $\Phi(p) = q$. Hence if there is a point p with $\Phi(p) = p$, the automorphism must agree with the identity automorphism.

We can now prove the theorem on equivalence of e-structures.

Proof. Let

$$h_U(p) = (y^1, \cdots, y^\rho, y^{\rho+1}, \cdots, y^n)$$

and $h_V(q) = (Y^1, \cdots, Y^\rho, Y^{\rho+1}, \cdots, Y^n)$, and let us fix the range of indices $1 \le \alpha, \beta, \gamma \le \rho$; $\rho + 1 \le a, b, c \le n$; $1 \le i, j, k \le n$. Then the regularity hypothesis implies

$$dy^\alpha = \sum y^\alpha_{|j}(y^1, \cdots, y^\rho)\omega^j_U$$

defines a $\rho \times n$ matrix $y^\alpha_{|j}$ of rank ρ. Relabel if necessary so that the first $\rho \times \rho$ block is nonsingular. (Note this will change the lexicographic ordering of $F_j(\omega_U)$ and $F_j(\Omega_V)$.)

Now

$$y^\alpha_{|j} = \left(y^\alpha_{|\beta} \mid y^\alpha_{|a}\right) \quad \text{with } \det y^\alpha_{|\beta} \neq 0$$

and let $(y^\alpha_{|\beta})^{-1} = g_{\alpha\beta}$. Then

$$\sum g_{\alpha\beta} dy^\beta = \omega^\alpha + \sum_{a=\rho+1}^{n} b_{\alpha a}\omega^a$$

or

$$\omega^\alpha = \sum a_{\alpha\beta} dy^\beta + \sum_{a=\rho+1}^{n} b_{\alpha a}\omega^a.$$

Similarly we have

$$G_{\alpha\beta} = (Y^\alpha_{;\beta})^{-1}, \qquad \Omega^\alpha = \sum A_{\alpha\beta} dY^\beta + \sum_{a=\rho+1}^{n} B_{\alpha a}\Omega^a.$$

Since

$$Y^\alpha_{;\beta} \circ \sigma = y^\alpha_{|\beta}$$

Cramer's rule guarantees

$$G_{\alpha\beta} \circ \sigma = g_{\alpha\beta},$$

and hence

$$A_{\alpha\beta} \circ \sigma = a_{\alpha\beta}, \qquad B_{\alpha a} \circ \sigma = b_{\alpha a}.$$

From

$$dy^\alpha = \sum y^\alpha_{|\beta}\omega^\beta + \sum y^\alpha_{|a}\omega^a$$

we see

$$dy^\alpha \wedge \omega^{\rho+1} \wedge \cdots \wedge \omega^n = \sum y^\alpha_{|\beta}\omega^\beta \wedge \omega^{\rho+1} \wedge \cdots \wedge \omega^n$$

and

$$dy^1 \wedge \cdots \wedge dy^\rho \wedge \omega^{\rho+1} \wedge \cdots \wedge \omega^n = \det(y^\alpha_{|\beta})\omega^1 \wedge \cdots \omega^\rho \wedge \omega^{\rho+1} \wedge \cdots \wedge \omega^n \neq 0.$$

Hence we have new coframings

$$\{dy^1, \cdots, dy^\rho, \omega^{\rho+1}, \cdots, \omega^n\} \quad \text{on } U,$$

and

$$\{dY^1, \cdots, dY^\rho, \Omega^{\rho+1}, \cdots, \Omega^n\} \quad \text{on } V,$$

which must be preserved by a diffeomorphism

$$\Phi : U \to V$$

solving the problem.

The technique of the graph cannot be applied in the usual form even if the system

$$\Delta_{U\times V}$$

were completely integrable, because it would produce maps Φ with

$$\Phi^* dY^\alpha = dy^\alpha$$

or

$$Y^\alpha \circ \Phi = y^\alpha + \text{constant}$$

and we need

$$Y^\alpha \circ \Phi = y^\alpha$$

on the nose.

To guarantee this, we must modify the technique of the graph by restricting to $\Sigma_{2n-\rho}$ the submanifold of $U \times V$ defined by setting

$$y^1 = Y^1, \cdots, y^\rho = Y^\rho$$

and considering the differential system

$$\pi_U^*\omega^a - \pi_V^*\Omega^a \quad \text{on } \Sigma_{2n-\rho}.$$

On $U \times V$

$$d(\pi_U^*\omega^a - \pi_V^*\Omega^a) = \sum \gamma_{jk}^a \circ \pi_U \, \pi_U^*\omega^j \wedge \pi_U^*\omega^k - \sum \Gamma_{jk}^a \circ \pi_V \, \pi_V^*\Omega^j \wedge \pi_V^*\Omega^k.$$

The hypothesis gives

$$\Gamma_{jk}^a \circ \sigma = \gamma_{jk}^a \quad \text{or} \quad \Gamma_{jk}^a \circ h_V^{-1} \circ h_U = \gamma_{jk}^a \quad \text{or} \quad \Gamma_{jk}^a \circ h_V^{-1} = \gamma_{jk}^a \circ h_U^{-1}.$$

Given $p \in U$ and $q \in V$ we have

$$\begin{aligned}\Gamma_{jk}^a(q) &= \Gamma_{jk}^a \circ h_U^{-1} \circ h_V(q) = \Gamma_{jk}^a \circ h_V^{-1}(Y^1(q), \cdots, Y^n(q)) \\ &= \Gamma_{jk}^a \circ h_V^{-1}(Y^1(q), \cdots, Y^n(q))\end{aligned}$$

and

$$\begin{aligned}\gamma_{jk}^a(p) &= \gamma_{jk}^a \circ h_V^{-1} \circ h_V(p) = \gamma_{jk}^a \circ h_V^{-1}(y^1(p), \cdots, y^n(p)) \\ &= \gamma_{jk}^a \circ h_V^{-1}(y^1(p), \cdots, y^n(p))\end{aligned}$$

hence on $\Sigma_{2n-\rho}$ we have

$$\gamma_{jk}^a \circ \pi_U = \Gamma_{jk}^a \circ \pi_V.$$

Now let us drop the π_U, π_V maps and show that the system $\omega^a - \Omega^a$ on $\Sigma_{2n-\rho}$ is completely integrable.

$$\begin{aligned}d(\omega^a - \Omega^a) &= \sum \gamma_{jk}^a(\omega^j \wedge \omega^k - \Omega^j \wedge \Omega^k) \\ &= \sum \gamma_{jk}^a\{\omega^j \wedge (\omega^k - \Omega^k) - \Omega^k \wedge (\omega^j - \Omega^j)\} \\ &= \sum \gamma_{j\alpha}^a \omega^\gamma \wedge (\omega^\alpha - \Omega^\alpha) + \sum \gamma_{jc}^a \omega^\alpha \wedge (\omega^c - \Omega^c) \\ &\quad - \sum \gamma_{\alpha k}^a \Omega^k \wedge (\omega^\alpha - \Omega^\alpha) - \sum \gamma_{ck}^a \Omega^k \wedge (\omega^c - \Omega^c) \\ &\equiv \sum (\gamma_{j\alpha}^a \omega^j - \gamma_{\alpha k}^a \Omega^k) \wedge (\omega^\alpha - \Omega^\alpha) \mod \{\omega^a - \Omega^a\}.\end{aligned}$$

Now a similar argument to that given above yields

$$A_{\alpha\beta} \circ \pi_V = a_{\alpha\beta} \circ \pi_U, \qquad B_{\alpha a} \circ \pi_V = b_{\alpha a} \circ \pi_U,$$

hence

$$\begin{aligned}\pi_U^*\omega^\alpha - \pi_V^*\Omega^\alpha &= \sum a_{\alpha\beta} \circ \pi_U dy^\beta + \sum b_{\alpha a} \circ \pi_U \pi_U^*\omega^a \\ &\quad - \sum A_{\alpha\beta} \circ \pi_V dY^\beta + \sum B_{\alpha a} \circ \pi_V \pi_V^*\Omega^a \\ &= \sum a_{\alpha\beta} \circ \pi_V(dy^\beta - dY^\beta) + \sum b_{\alpha a} \circ \pi_V(\pi_U^*\omega^a - \pi_V^*\Omega^a).\end{aligned}$$

Additionally, on Σ_{2n-p}, we have $dy^\beta = dY^\beta$, $\omega^\alpha - \Omega^\alpha \equiv 0 \bmod \{\omega^a - \Omega^a\}$ and the previous calculation gives

$$d(\omega^a - \Omega^a) \equiv 0 \quad \mathrm{mod}\{\omega^a - \Omega^a\},$$

which proves the integrability as needed. A unique leaf of $\{\omega^a - \Omega^a\}$ through a point

$$(y^1, \cdots, y^\rho, y^{\rho+1}, \cdots, y^n; Y^1, \cdots, Y^\rho, Y^{\rho+1}, \cdots, Y^n)$$

in Σ_{2n-p} gives a submanifold $\mathcal{L}$ of dimension equal to $2n - p - (n-p) = n$. This leaf is an integral manifold of the differential system

$$\{dy^1 - dY^1, \cdots, dy^\rho - dY^\rho, \omega^{\rho+1} - \Omega_{\rho+1}, \cdots, \omega^n - \Omega^n\},$$

and the arguments given in the general technique of the graph imply there exists $\Phi : U \to V$ with

$$Y^\alpha \circ \Phi = y^\alpha \quad \text{and} \quad \Phi^*\Omega^a = \omega^a.$$

A repetition of the arguments given above yields

$$A_{\alpha\beta} \circ \Phi = a_{\alpha\beta}, \qquad B_{\alpha a} \circ \Phi = b_{\alpha a}$$

and as a result

$$\begin{aligned} \Phi^*\Omega^\alpha &= \sum A_{\alpha\beta} \circ \Phi dY^\beta \circ \Phi + B_{\alpha a} \circ \Phi \Phi^*\Omega^a \\ &= \sum a_{\alpha\beta} dY^\beta + \sum b_{\alpha a}\omega^a \\ &= \omega^\alpha \end{aligned}$$

as desired.

Note that the condition

$$F_{j+1}(\Omega) \circ \sigma = F_{j+1}(\omega)$$

is actually fairly complicated to check. Every function in $F_{j+1}(\Omega)$ depends only on

$$(Y^1, \cdots, Y^\rho)$$

and every function in $F_{j+1}(\omega)$ depends only on

$$(y^1, \cdots, y^\rho).$$

Thus the fact that σ takes each element of the lexicographic set $F_{j+1}(\Omega)$ onto the corresponding element of the identically constructed lexicographic set $F_{j+1}(\omega)$ is a statement about functions of $(Y^1, \cdots, Y^\rho)$ and $(y^1, \cdots, y^\rho)$; hence this means that all the dependencies of functions on $F_{j+1}(\Omega)$ must be taken into identical dependencies of the corresponding functions in $F_{j+1}(\omega)$. The next example should help make the meaning of these conditions clear.

An important geometric consequence of this theorem is the following.

COROLLARY. *Let ω be a regular e-structure at p of rank ρ, then the orbits of the automorphisms of ω locally define an $(n-p)$-dimensional foliation in a neighborhood of p.*

Proof. The foliation is defined by the differential system $(dy^1, \cdots, dy^\rho)$, and the proof of the general theorem shows that each leaf

$$y^1 = c^1, \cdots, y^\rho = c^\rho$$

with $c^1, \cdots, c^\rho$ constant is the orbit of the pseudogroup of local equivalences.

In particular, we note the following corollary.

COROLLARY. *Let ω be a regular e-structure with $\rho = 0$, i.e., constant torsion, then the automorphisms are locally transitive.*

Although we have solved the equivalence problem for the case where there is an e-structure on $U \times G$, it is unpleasant to have the group variables involved in the invariants. It is often possible to continue the reduction process to produce an e-structure on U. This is possible since G may still act nontrivially on the new reduced torsion even if $\mathcal{G}^{(1)} = 0$.

Example 3. *Web geometry (continued).* We had deduced structure equations on $U \times G$

$$\begin{aligned} d\omega^1 &= \varphi \wedge \omega^1 \\ d\omega^2 &= \varphi \wedge \omega^2 \\ d\varphi &= c\,\omega^1 \wedge \omega^2 \end{aligned}$$

with $\omega^1, \omega^2, \varphi$ unique. Hence we have an e-structure on $U \times G$. Differentiation of the last equation gives

$$0 = d^2\varphi = dc \wedge \omega^1 \wedge \omega^2 + c\,\varphi \wedge \omega^1 \wedge \omega^2 - c\,\varphi \wedge \omega^2 \wedge \omega^1$$

or

$$dc \equiv -2c\,\varphi \mod (\omega^1, \omega^2).$$

We have already disposed of the case $c = 0$, hence we assume $c \neq 0$ so that the group G acts on c multiplicatively by the inverse of a square. This means that the reduction process could be carried out one step further to make $c = \pm 1$. Here there appear to be two orbits, but we can consider them together, since the case $d\varphi = -\omega^1 \wedge \omega^2$ is reduced to the case $d\varphi = \omega^1 \wedge \omega^2$ by interchanging ω^1 and ω^2. The reduction

$$\varphi \equiv 0 \mod (\omega^1, \omega^2),$$

reduces the group to the identity and produces an e-structure on U. If we let

$$\varphi = B\,\omega^1 - A\,\omega^2$$

we have

$$d\omega^1 = A\,\omega^1 \wedge \omega^2, \qquad d\omega^2 = B\,\omega^1 \wedge \omega^2.$$

These coefficients are not arbitrary, if we let A_1, A_2, B_1, B_2 be the first covariant derivatives with respect to ω_1 and ω_2, differentiation of φ gives

$$d\varphi = -(B_2 + A_1)\omega^1 \wedge \omega^2 + (BA - AB)\omega^1 \wedge \omega^2,$$

but we also have the structure equation

$$d\varphi = \omega^1 \wedge \omega^2,$$

hence

$$A_1 + B_2 = -1.$$

In particular A_1 and B_2 are not both zero, which means A and B cannot both be constants. The converse to the first Lie theorem then implies that there is no two-dimensional symmetry group of this system. This means the rank cannot be zero and is at most two. If it is rank 2, then the equivalence theorem for e-structures says the equivalences are unique, hence let us assume the rank is one. Then

$$F_2(\omega) = \{A, B, A_1, A_2, B_1, B_2\}$$

are all functions of a single parameter τ. Hence by the straightforward application of the theorem we would need to check that five dependencies are preserved. It is of course possible to work harder and reduce the number of dependencies needed. In this case we can get down to one dependency by choosing τ more subtly than the direct construction given in the theorem. Since $A_1 + B_2 = -1$ either A_1 or B_2 is not zero, but let us assume $A_1 \neq 0$.

Exercise. $A_1 \neq 0$ implies $A_2 \neq 0$.

Now we have

$$\frac{dA}{A_1} = \omega^1 + \frac{A_2}{A_1}\omega^2$$

and

$$\frac{dA}{A_1} = f(\tau)\,d\tau \neq 0, \qquad \frac{A_2}{A_1} = g(\tau) \neq 0.$$

If we choose a new parameter t such that $t = g(\tau)$, then we have a function $F(t)$ defined by

$$dt = F(t)(\omega^1 + t\,\omega^2).$$

Exercise. Show

$$F(t) = -A - tB, \qquad A = tF - F - t/F, \qquad B = 1/F - F'.$$

As a result of the exercise the necessary and sufficient conditions are reduced to requiring that the dependence of

$$-A - A_2/A_1 B \quad \text{on } A_2/A_1$$

is the same. This is because all the covariant derivatives of A and B can be expressed in terms of $t, F(t)$ and their derivatives.

The set of differential equations admitting a one-parameter group of symmetries can be determined by studying its infinitesimal generators.

Let X be the infinitesimal generator of a one-parameter transformation group of symmetries of the differential equations, then

$$L_X\omega^1 = 0, \qquad L_X\omega^2 = 0.$$

Let $\{e_1, e_2\}$ be dual basis to ω^1, ω^2 and let

$$X = \lambda_1 e_1 + \lambda_2 e_2 = X_1 + X_2$$

then $\langle X, \omega^1\rangle = \lambda_1$, $\langle X, \omega^2\rangle = \lambda_2$. Now

$$\begin{aligned} 0 = L_X\omega^1 &= X \lrcorner d\omega^1 + d(X \lrcorner \omega^1) = A\lambda_1\omega^2 - A\lambda_2\omega^1 \\ &= d\lambda_1 - A(\lambda_2\omega^1 - \lambda_1\omega^2). \end{aligned}$$

Similarly,

$$0 = d\lambda_1 - B(\lambda_1\omega^1 - \lambda_1\omega^2).$$

Next we see

$$\begin{aligned} \langle [X_1, X_2], \omega^1\rangle &= -\langle X_1 \wedge X_2, d\omega^1\rangle + X_1\langle X_2, \omega^1\rangle - X_2\langle X_1, \omega^1\rangle \\ &= -A \det\begin{pmatrix} \langle X_1, \omega^1\rangle & \langle X_1, \omega^1\rangle \\ \langle X_2, \omega^1\rangle & \langle X_2, \omega^2\rangle \end{pmatrix} + X_1(0) - X_2(\lambda_1) \\ &= -A \det\begin{pmatrix} \lambda_1 & 0 \\ 0 & \lambda_2 \end{pmatrix} - \langle X_2, d\lambda_1\rangle \\ &= -(A\lambda_1\lambda_2) - (-A\lambda_1\lambda_2) \\ &= 0 \end{aligned}$$

and similarly $\langle [X_1, X_2], \omega^2\rangle = 0$. As such by the Frobenius theorem we can write

$$\frac{\partial}{\partial X} = X_1 \quad \text{and} \quad -\frac{\partial}{\partial Y} = X_2$$

which creates the normal form $X = \partial/\partial X - \partial/\partial Y$. Now we have seen $\omega^1 = g(X, Y)dX$ and $\omega^2 = h(X, Y)dY$ and since

$$\begin{aligned} L_X\omega^1 &= X \lrcorner (dg \wedge dX) + d(X \lrcorner \omega^1) \\ &= X(g)dX - dg + dg \\ &= X(g)dX \end{aligned}$$

we see that $L_X\omega^1 = 0$ and $L_X\omega^2 = 0$ if and only if

$$g(X, Y) = g(X + Y) \quad \text{and} \quad h(X, Y) = h(X + Y).$$

As such,

$$\begin{aligned} \omega^2 - \omega^1 &= h(X + Y)dY - g(X + Y)dX \\ &= h(X + Y)\left\{dY - \frac{g(X + Y)}{h(X + Y)}dX\right\} \end{aligned}$$

and setting $U(X+Y) = g(X+Y)/h(X+Y)$ we see these equations are those equivalent to the ones of the form $dY/dX = U(X+Y)$.

Example 1. *Riemannian geometry* (*continued*). At the end of Lecture 3 we saw that there was a unique vector 1-form ω and a matrix 1-form ϕ on $U \times G$. In order to use the theorem on equivalence of e-structures we need to analyze ϕ and $d\phi$.

We know $\phi = dSS^{-1} + \theta_U(u,S)$ where $\theta_U(u,S)$ is semibasic. Since ϕ is a form of type adjoint, that is, left translation L_C on G satisfies

$$L_C^*\phi = C\phi C^{-1}, \tag{10}$$

we deduce

$$L_C^*\theta_U(u,CS) = C\,\theta_U(u,S)\,C^{-1}.$$

Exercise. Use the structure equation $d\omega = \phi \wedge \omega$ to verify (10). Let $C = S^{-1}$ and we have

$$L_{S^{-1}}^*\theta_U(u,e) = S^{-1}\theta_U(u,S)S = \theta_U(u,e)$$

or

$$\theta_U(u,S) = S\theta_U(u)S^{-1}.$$

The coefficients Γ_{ik}^j defined by

$$(\theta_U)_i^j = \sum \Gamma_{ik}^j \omega^k$$

are called the Christoffel symbols of the Levi-Civita connection. (Not to be confused with other Γ_{ik}^j in this report, but giving these a different name would be like letting ϵ be less than zero.) Applying d to the first structure equation, we find

$$\begin{aligned} 0 = d(d\omega) &= d\phi \wedge \omega - \phi \wedge d\omega \\ &= (d\phi - \phi \wedge \phi) \wedge \omega. \end{aligned}$$

Let

$$\Theta = d\phi - \phi \wedge \phi,$$

then using indices we see

$$0 = \sum \Theta_j^i \wedge \omega^j, \quad \text{where } \Theta_j^i = -\Theta_i^j$$

and hence

$$\Theta_j^i = \sum \psi_{jk}^i \wedge \omega^k.$$

Substitution back into the last equation gives

$$0 = \sum \omega^j \wedge \psi_{jk}^i \wedge \omega^k,$$

and wedging with the $(n-2)$-forms consisting of the wedge products of $(n-2)$ omegas gives

$$0 = \omega^1 \wedge \cdots \wedge \omega^n \wedge (\psi^i_{jk} - \psi^j_{ik}).$$

Similarly we can wedge the defining equation of ψ^i_{jk} with the $(n-1)$-forms consisting of the wedge products of $(n-1)$ omegas and use the skew symmetry of the Θ^j_i to find

$$\omega^1 \wedge \cdots \wedge \omega^n \wedge \psi^j_{ik} = -\omega^1 \wedge \cdots \wedge \omega^n \wedge \psi^i_{jk}.$$

This means

$$\begin{aligned} \psi^i_{jk} &\equiv -\psi^j_{ik} \\ \psi^i_{jk} &\equiv \psi^i_{kj} \end{aligned} \quad \mathrm{mod}\,(\omega^1, \cdots, \omega^n)$$

and hence by the S_3-lemma

$$\psi^i_{jk} \equiv 0 \quad \mathrm{mod}\,(\omega^1, \cdots, \omega^n).$$

As such we see that the ψ^i_{jk} are semibasic and we may define coefficients

$$\psi^i_{jk} = \frac{1}{2} \sum S^i_{jkl}\, \omega^l, \quad \text{with } S^i_{jkl} = -S^i_{jlk}.$$

Exercise. Use L^*_C to show $\Theta_U(u, S) = S\Theta_U(u, e)S^{-1}$.
As a result of the exercise we may define coefficients

$$\Theta_U(u, e) = \frac{1}{2} \sum R^i_{jkl} \omega^l, \quad \text{with } R^i_{jkl} = -R^i_{jlk}.$$

which are the components of the Riemann–Christoffel curvature tensor. This allows us to state the solution of the *Christoffel form problem* (see [Chr. 1869]), which in its classical, folkloric form goes something like this.

THEOREM. *A Riemannian metric is determined up to isometries by prescribing the Christoffel symbols, the Riemannian curvature tensor and its derivatives up to order* $n+1$.

Note that the conditions of regularity, equal order, and equal rank, as well as preservation of dependencies, are classically not stated. *It is necessary to be very careful in reading classical results on equivalence.* The precise statement is a repeat of the *e*-structure theorem in this special case.

LECTURE 7

Global Results and Involutive Structures

Let M be a manifold with an e-structure ω. Introducing the structure functions

$$d\omega^i = \frac{1}{2}\sum c^i_{jk}(u)\omega^j \wedge \omega^k, \qquad c^i_{jk} = -c^i_{kj}$$

we say that an *e-structure is integrable* if

$$dc^i_{jk} = 0,$$

i.e., the structure functions are all constant.

As we have seen, a Lie group is an example of an integrable e-structure. We will show that under suitable restrictions the converse is also true.

The idea is to characterize the candidates for right translation. If M were a Lie group, let $T_{ab} : M \to M$ be the right translation satisfying $T_{ab}(a) = b$, i.e., $T_{ab} = R_{a^{-1}b}$. As we saw in Lecture 3, T_{ab} is characterized by the solution of the initial value problem

$$T^*_{ab}\omega|_{T_{ab}(s)} = \omega|_s, \qquad T_{ab}(a) = b.$$

We want to understand the construction of $T_{ab}(x)$ precisely so we can try to mimic the group construction in the case of an e-structure. Naturally we try to use the technique of the graph.

Now

$$T_{ab}(x) = R_{a^{-1}b}(x)$$

is equivalent to

$$T_{ab}(x)b^{-1} = xa^{-1}$$

or

$$(x, T_{ab}(x)) \in \Delta(a, b),$$

where Δ is the diagonal subgroup. This means $(x, T_{ab}(x))$ lies in the leaf of $\Delta_{G\times G}$ passing through (a, b).

This gives the following procedure to try on a general integrable e-structure ω. Let π_L and π_R be the natural projections on $M \times M$ to M and let

$$\tilde{\omega} = \pi^*_L\omega \quad \text{and} \quad \Omega = \pi^*_R\omega.$$

Then as in the technique of the graph we define an exterior differential system

$$\Delta_{M\times M} = \tilde{\omega} - \Omega,$$

which satisfies the Frobenius conditions since ω is integrable and has the diagonal $\Delta : M \to M \times M$ given by $\Delta(x) = (x, x)$ as an integral manifold.

Now given $a, b \in M$ choose the leaf $\mathcal{L}$ of $\Delta_{M\times M}$ through (a, b); furthermore, given $x \in M$ look at the inverse image of x in $\mathcal{L}$, this results in a point $(x, T_{ab}(x))$ which defines $T_{ab}(x)$.

This is the idea, but to be well defined and a diffeomorphism,

$$\pi_L : \mathcal{L} \to M$$

must be a simply connected covering. There are various criteria to guarantee that this will happen. One of the most important is the following.

Given an e-structure ω, there is a natural Riemannian metric

$$ds^2 = \sum (\omega^i)^2.$$

We say the *e-structure is complete* if the manifold is metrically complete.

Under the assumption of completeness, Hopf–Rinow type analysis of the leaf $\mathcal{L}$ can be shown to evenly cover M. The added hypothesis that M is simply connected gives the necessary conditions for this construction.

THEOREM. *Let M be a connected, simply connected manifold with a complete integrable e-structure. Then M is a Lie group.*

Proof. Choose $x_0 \in M$ and let the identity be defined by

$$R_e = T_{x_0 x_0}.$$

Then for any $x \in M$, $R_e x$ is obtained by taking the leaf through (x_0, x_0) and looking at the inverse image of x, but this leaf is the diagonal Δ. Hence the inverse image is (x, x) which shows

$$R_e(x) = x.$$

Given any $a \in M$ we define

$$R_a = T_{x_0 a}, \qquad R_{a^{-1}} = (T_{x_0 a})^{-1}.$$

Clearly

$$R_a \circ R_{a^{-1}} = Id = R_{a^{-1}} \circ R_a.$$

The associative law is equivalent to showing

$$R_{R_c(a)}(b) = R_c \circ R_a(b),$$

but

$$f(x) = R_{R_c(a)}(x) \quad \text{and} \quad g(x) = R_c \circ R_a(x)$$

are automorphisms, and since

$$f(x_0) = R_{R_c(a)}(x_0) = T_{x_0 T_{x_0 c}(a)}(x_0) = T_{x_0 c}(a),$$
$$g(x_0) = R_c \circ R_a(x_0) = T_{x_0 c}(T_{x_0 a}(x_0)) = T_{x_0 c}(a)$$

they agree at x_0. If we let

$$h(x) = g^{-1}(f(x))$$

we see

$$h(x_0) = x_0,$$

hence by the corollary to the technique of the graph in Lecture 6 on fixed points of automorphisms we see $h =$ identity and $f = g$.

Understanding when the hypotheses on the leaf $\mathcal{L}$ hold is a delicate business. For example, there are discrete arithmetic subgroups Γ of $Sl(2, \mathbf{R})$ with compact quotients such that the first Betti number

$$b_1(Sl(2, \mathbf{R})/\Gamma) = 1.$$

Now $Sl(2, \mathbf{R})/\Gamma$ inherits an e-structure from the Maurer–Cartan forms on $Sl(2, \mathbf{R})$, but if it were a Lie group, it would be simple since its Lie algebra would be the same as $Sl(2, \mathbf{R})$. This would violate the first Whitehead theorem that the first Betti number of a compact simple Lie group is zero.

Quasilinear systems in involution.

Now we take up the case when all the normalizations and reductions are completed, but group parameters remain. The procedure is simple but is based on powerful mathematics with awesome results.

The structure equations

$$d\omega^i = \sum a^i_{jp} \pi^p \wedge \omega^j + \frac{1}{2} \sum \gamma^i_{jk} \omega^j \wedge \omega^k$$

are of the form of a quasilinear exterior differential system. This is a class of exterior differential systems for which the Cartan–Kähler theorem is very easy to apply. The Cartan–Kähler theorem is a geometric formulation and extension of the Cauchy–Kowaleski existence theorem in partial differential equations, and it applies when an exterior differential system is analytic and in involution. (This is not Chevalley's simple notion of involution which appears in the elementary Frobenius theorem.) We will now formulate Cartan's test for a quasilinear system to be in involution. (See [Gr-J. 1987] for an overview or [BC3G. 1989] for the full story.)

The matrix

$$(\pi^j_i) = \left(\sum a^j_{ip} \pi^p\right)$$

is called the tableau matrix and

$$(\pi_i^j) \quad \text{mod} \quad (\omega^1, \cdots, \omega^m)$$

is called the reduced tableau matrix. Next we give an inductive definition of the *reduced Cartan characters* $\sigma_1, \cdots, \sigma_p$.

The induction is started by $\Sigma_0 = \{0\}$. Now choose as many independent entries mod $\Sigma_0, \cdots, \Sigma_{k-1}$ as possible by taking at most one from each row of (π_i^j) mod ω. The number of rows with a nonzero contribution is called σ_k, and the collection of independent forms is called Σ_k.

For example, the reduced tableau matrix with all entries independent,

$$\left(\pi_i^j\right) = \begin{pmatrix} 0 & \alpha_1 & 0 \\ 0 & \alpha_2 & 0 \\ \beta_1 & \beta_2 & \gamma \end{pmatrix},$$

has

$$\sigma_1 = 3, \qquad \sigma_2 = 1, \qquad \sigma_3 = 1,$$

and

$$\Sigma_1 = \{\alpha_1, \alpha_2, \beta_1\}, \quad \Sigma_2 = \{\beta_2\}, \quad \Sigma_3 = \{\gamma\}.$$

Note the Σ_i are not unique.

As usual we let $\mathcal{G}^{(1)}$ be the kernel of the mapping L. Then it is known that

$$\dim \mathcal{G}^{(1)} \leq \sigma_1 + 2\sigma_2 + \cdots + p\sigma_p.$$

Recall that by (9) in Lecture 4, $\dim \mathcal{G}^{(1)}$ is the number of parameters of indeterminancy in the forms in the reduced tableau.

Let σ be defined as the sum on the right-hand side. The tableau is said to be *in involution* or *involutive* if

$$\dim \mathcal{G}^{(1)} = \sigma.$$

This is really taking a basic theorem, called Cartan's test, and in the finest Bourbaki tradition, making it a definition. We now state the General Equivalence Theorem of Cartan.

THEOREM. *Let $U \subseteq \mathbf{R}^{n+r}$ be an open set. Let $x = (x^\alpha) : U \longrightarrow \mathbf{R}^m$ be a submersion, and let $\omega^1, \cdots, \omega^n$ be linearly independent 1-forms on U satisfying the following hypotheses:*

(1) *There exist functions J_i^α on $x(U) \subseteqq \mathbf{R}^m$ so that*

$$dx^\alpha = J_i^\alpha(x)\, \omega^i;$$

(2) *There exist 1-forms $\pi^1, \cdots, \pi^r$ on U independent from $\omega^1, \cdots, \omega^1, \cdots, \omega^n$ and functions a_{jp}^i, $c_{jk}^i = -c_{kj}^i$ on $x(U) \subseteq \mathbf{R}^m$ so that*

$$d\omega^i = -\sum a_{jp}^i \pi^p \wedge \omega^j + \frac{1}{2} \sum c_{jk}^i\, \omega^j \wedge \omega^k,$$

where for each $x_0 \in x(U)$ the tableau $a^i_{j\rho}(x_0)$ is involutive for each $x_0 \in x(U)$. Moreover, the characters $\sigma_i(x_0)$ are constant (i.e., independent of x_0);

(3) *The ω^i are real analytic in some coordinate system $x^1, \cdots, x^m, y^1, \cdots, y^{n+r}$ on U. Then for any two points $u_1, u_2 \in U$ with $x(u_1) = x(u_2)$ there exist open sets $V_1, V_2 \subseteq U$ with $u_i \in V_i$ and a diffeomorphism $f : V_1 \longrightarrow V_2$ satisfying*

$$f(u_1) = u_2, \qquad f^*(\omega^i) = \omega^i, \qquad f^*(x^\alpha) = x^\alpha.$$

This theorem is an elementary consequence of the Cartan–Kähler theorem. The proof is given by simply letting

$$X = \{(u_1, u_2) \in U | x(u_1) = x(u_2)\}$$

and constructing a differential system on X as follows: Let $\pi_1, \pi_2 : X \longrightarrow U$ be the projections onto the first and second factors. Set

$$\theta^i = \pi_2^*(\omega^i) - \pi_1^*(\omega^i)$$

and continue to denote $\pi_1^*(x^\alpha)$ $(= \pi_2^*(x^\alpha))$ by x^α. Note that only $n - m$ of the θ^i are independent:

$$\begin{aligned} J\theta &= J(\pi_2^*(\omega) - \pi_1^*(\omega)) = \pi_2^*(J\omega) - \pi_1^*(J\omega) \\ &= dx - dx = 0. \end{aligned}$$

We consider as independence conditions the forms $\pi_1^*(\Omega), \pi_2^*(\Omega)$, where

$$\Omega = \omega^1 \wedge \cdots \wedge \omega^n.$$

By hypothesis, if I is the differential system generated by $\theta^i, d\theta^i$, then I is involutive with respect to either $\pi_1^*(\Omega)$ or $\pi_2^*(\Omega)$. Since there clearly exist integral elements satisfying both independence conditions simultaneously, we are done: an integral manifold of (I, π_1^*) passing through (u_1, u_2) is the graph of a solution f.

If the tableau a^i_{jp} is trivial (i.e., $a^i_{jp} = 0$ and $r = 0$), then the assumption of real analyticity is not needed. In this case, the theorem is just a generalization of the third fundamental theorem of Lie.

If the characteristic variety of the tableau a^i_{jp} is appropriately "hyperbolic" then analyticity is also not needed. In particular, if the characteristic variety consists of real distinct points, then analyticity is not needed.

Cartan derives necessary and sufficient conditions on functions $(a^i_{jp}, c^i_{jk}, J^\alpha_i)$ defined on an open set in $\mathbf{R}^m$ that there exist forms ω^i and a submersion on $U \subseteq \mathbf{R}^{n+r}$

$$x : U \longrightarrow \mathbf{R}^m$$

satisfying the conditions of the theorem. This is the true generalization of the third fundamental theorem of Lie (see [Si–S. 1965]).

Thus the action of local equivalences is transitive on this set of equivalence problems. This is usually described by saying the problem admits a transitive pseudogroup. In this case the final move is to find a single, natural, and simple example with the given structure equations. This then is a normal form for this class of problems.

Example 2. *Conformal geometry* (*continued*). The group $CO(n,\mathbf{R})$ has Maurer–Cartan matrix

$$\begin{aligned} d(\lambda S)(\lambda S)^{-1} &= d\lambda SS^{-1}\lambda^{-1} + \lambda dSS^{-1}\lambda^{-1} \\ &= \frac{d\lambda}{\lambda}I + dSS^{-1}, \end{aligned}$$

where $S \in SO(n,\mathbf{R})$.

If we consider the case $n = 2$ we have the Maurer–Cartan matrix

$$\begin{pmatrix} d\lambda/\lambda & \varphi \\ -\varphi & d\lambda/\lambda \end{pmatrix};$$

hence if

$$d\begin{pmatrix} \omega^1 \\ \omega^2 \end{pmatrix} = \begin{pmatrix} \alpha & \beta \\ \gamma & \delta \end{pmatrix} \wedge \begin{pmatrix} \omega^1 \\ \omega^2 \end{pmatrix}$$

we have principal components

$$\alpha - \delta \quad \text{and} \quad \beta + \gamma.$$

As a result we can write

$$d\begin{pmatrix} \omega^1 \\ \omega^2 \end{pmatrix} = \begin{pmatrix} \alpha & \beta \\ -\beta & \alpha \end{pmatrix} \wedge \begin{pmatrix} \omega^1 \\ \omega^2 \end{pmatrix} + \begin{pmatrix} A \\ B \end{pmatrix} \omega^1 \wedge \omega^2$$

and absorb all the torsion giving

$$d\begin{pmatrix} \omega^1 \\ \omega^2 \end{pmatrix} = \begin{pmatrix} \alpha & \beta \\ -\beta & \alpha \end{pmatrix} \wedge \begin{pmatrix} \omega^1 \\ \omega^2 \end{pmatrix}.$$

Since no more reduction is possible we check for involution. The reduced Cartan characters are $\sigma_1 = 2$ and $\dim \mathcal{G}^{(1)}$ is given by the number of parameters of indeterminancy. The equation for $d\omega^1$ and Cartan's lemma show that α and β are determined up to three parameters, but the equation for $d\omega^2$ and Cartan's lemma show that two of these parameters must be equal; hence $\dim \mathcal{G}^{(1)} = 2$ and the system is in involution.

As a result there is a transitive pseudogroup and all analytic metrics in two dimensions are equivalent under conformal transformations. In particular, every analytic metric can be written in the form $ds^2 = \lambda^2((dx^1)^2 + (dx^2)^2)$. (This result is also true for C^∞ metrics with the result due to Korn–Lichtenstein (see [Ch. 1955]), although an estimate needs to be done more carefully.)

Example 9. *Second-order ordinary differential equations under contact transformations*. Choose

$$\omega_U^1 = dy - y'dx, \qquad \omega_U^2 = dy' - f\,dx, \qquad \omega_U^3 = dx$$

and note the integrability

$$d\omega_U^1 = -dy' \wedge dx = -(dy' - f\,dx) \wedge dx = -\omega_U^2 \wedge \omega_U^3.$$

The equivalence problem lifts to

$$\begin{pmatrix} \omega^1 \\ \omega^2 \\ \omega^3 \end{pmatrix} = \begin{pmatrix} a & 0 & 0 \\ b & c & 0 \\ e & g & h \end{pmatrix} \begin{pmatrix} \omega_U^1 \\ \omega_U^2 \\ \omega_U^3 \end{pmatrix}.$$

Note that the variable g is needed since $g = 0$ would give point transformations. After using the principal components and absorption the structure equations are

$$d\begin{pmatrix} \omega^1 \\ \omega^2 \\ \omega^3 \end{pmatrix} = \begin{pmatrix} \alpha & 0 & 0 \\ \beta & \gamma & 0 \\ \epsilon & \zeta & \xi \end{pmatrix} \wedge \begin{pmatrix} \omega^1 \\ \omega^2 \\ \omega^3 \end{pmatrix} + \begin{pmatrix} A \\ 0 \\ 0 \end{pmatrix} \omega^3 \wedge \omega^2.$$

Since

$$d\omega^1 \wedge \omega^1 = a^2 d\omega_U^1 \wedge \omega_U^1 = -a^2 \omega_U^1 \wedge \omega_U^2 \wedge \omega_U^3 \neq 0$$

and

$$d\omega^1 \wedge \omega^1 = -A\,\omega^1 \wedge \omega^2 \wedge \omega^3$$

we see $A \neq 0$.

Computing $d^2\omega^1 \bmod \omega^1$ we see

$$dA - \alpha A + A\xi + A\gamma \equiv 0 \quad \text{mod base},$$

hence we may normalize

$$A \to 1$$

which induces a principal component of order 2,

$$-\xi + \alpha - \gamma.$$

This leads to the structure equations

$$d\begin{pmatrix} \omega^1 \\ \omega^2 \\ \omega^3 \end{pmatrix} = \begin{pmatrix} \alpha & 0 & 0 \\ \beta & \gamma & 0 \\ \epsilon & \zeta & \alpha - \gamma \end{pmatrix} \wedge \begin{pmatrix} \omega^1 \\ \omega^2 \\ \omega^3 \end{pmatrix} + \begin{pmatrix} 1 \\ 0 \\ 0 \end{pmatrix} \omega^3 \wedge \omega^2$$

whose torsion terms have been absorbed into ϵ and ζ.

Since no more reduction is possible we check for involution. The reduced Cartan characters are

$$\sigma_1 = 3, \qquad \sigma_2 = 2,$$

hence σ equals

$$\sigma_1 + 2\sigma_2 = 3 + 4 = 7.$$

Then dim $\mathcal{G}^{(1)}$ is computed by finding the number of parameters of indeterminancy, which follows from multiple applications of Cartan's lemma

$$\alpha \text{ gives one; } \beta, \gamma \text{ give three; } \epsilon, \zeta \text{ give three,}$$

hence

$$\dim \mathcal{G}^{(1)} = 1 + 3 + 3 = 7,$$

and the system is in involution. As a result there is a transitive pseudogroup and all analytic second-order equations $y'' = f(x, y, y')$ are equivalent under contact transformations. In particular, we have Lie's theorem that all are contact equivalent to $y'' = 0$. By using the Pfaff problem the analyticity assumption in this theorem may be avoided.

What about the case in which the tableau is involutive, but the torsion is not constant? In this case there are pseudogroups of equivalences, but they are not transitive. The analysis here is difficult, because the isotropy groups of each orbit will depend on the torsion, and leads to an equivalence problem in which the group changes from point to point. Cartan indicates the procedure even in this case, but as we have said before, the complexity of having the $a^i_{j\rho}$ functions of an independent set of components of the torsion would only cloud the general method. Unfortunately, this case cannot be avoided since it does arise in interesting problems.

We can summarize this section with a slightly enlarged flowchart (see Fig. 2).

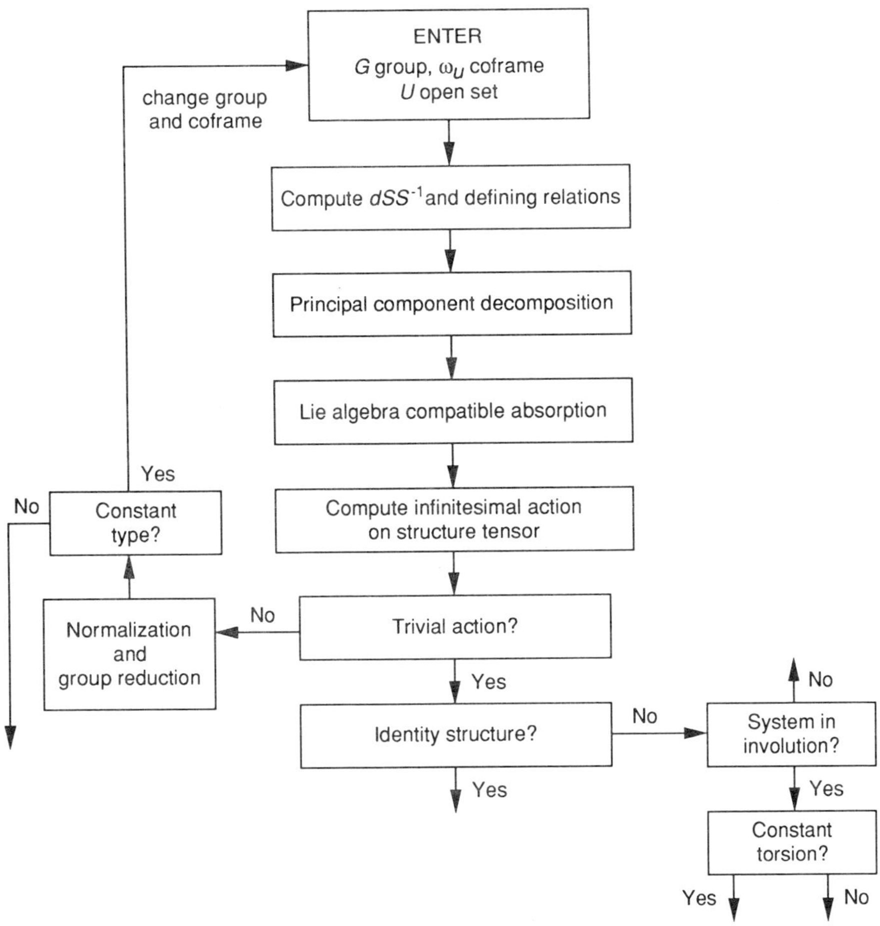

FIG. 2 *Flowchart.*

LECTURE 8

Serendipity

We now take up a major example, that of *state estimation of plants under feedback*. In standard language this is as follows.

Example 7. *Invariants of underdetermined systems of ordinary differential equations under diffeomorphisms of the form*

$$\Phi(t,x,u) = (t,\phi(x),\psi(x,u)), \quad \textit{where } x \in \mathbf{R}^m \textit{ and } u \in \mathbf{R}^p.$$

This is the *geometry of control systems under feedback*. The x variables are known as *states* and the u variables are known as *controls*.

Let (V,X,U) and (U,x,u) be open sets with coordinates in $\mathbf{R}^m \times \mathbf{R}^n$ and underdetermined systems of ordinary differential equations

$$\frac{dx}{dt} = f(x,u) \text{ and } \frac{dX}{dT} = F(X,U).$$

where we assume $f \neq 0$ and $F \neq 0$.

This is an overdetermined problem, and we apply the overdetermined algorithm from Lecture 5. Let us introduce on $U \subset \mathbf{R} \times \mathbf{R}^m \times \mathbf{R}^p$

$$\omega_U^1 = dt, \qquad \omega_U^2 = dx, \qquad \omega_U^3 = du,$$

a 1-form, an m-vector 1-form, and a p-vector 1-form with analogous forms defined by capital omegas on V. Then the feedback diffeomorphisms are characterized by

$$\Phi^* \begin{pmatrix} \Omega_V^1 \\ \Omega_V^2 \\ \Omega_V^3 \end{pmatrix} = \begin{pmatrix} 1 & 0 & 0 \\ 0 & A & 0 \\ 0 & B & C \end{pmatrix} \begin{pmatrix} \omega_U^1 \\ \omega_U^2 \\ \omega_U^3 \end{pmatrix}$$

which defines G.

In order that Φ preserve integral curves we must require

$$\Phi^*(dX - F\,dT) = a(dx - f\,dt), \qquad a \in Gl(m,\mathbf{R}).$$

By introducing a second coframe on U defined by

$$\theta_U^1 = dt, \qquad \theta_U^2 = dx - f\,dt, \qquad \theta_U^3 = du$$

with analogous forms denoted by capital thetas on V, we see that we also require

$$\Phi^* \begin{pmatrix} \Theta_V^1 \\ \Theta_V^2 \\ \Theta_V^3 \end{pmatrix} = \begin{pmatrix} 1 & 0 & 0 \\ 0 & a & 0 \\ e & b & c \end{pmatrix} \begin{pmatrix} \theta_U^1 \\ \theta_U^2 \\ \theta_U^3 \end{pmatrix}$$

which defines G'.

If we lift to $U \times G \times G'$ we have

$$\omega = \begin{pmatrix} 1 & 0 & 0 \\ 0 & A & 0 \\ 0 & B & C \end{pmatrix} \omega_U, \qquad \theta = \begin{pmatrix} 1 & 0 & 0 \\ 0 & a & 0 \\ e & b & c \end{pmatrix} \theta_U,$$

and since

$$\omega_U = \begin{pmatrix} 1 & 0 & 0 \\ f & 1 & 0 \\ 0 & 0 & 1 \end{pmatrix} \theta_U$$

we have

$$\begin{aligned} \omega &= \begin{pmatrix} 1 & 0 & 0 \\ 0 & A & 0 \\ 0 & B & C \end{pmatrix} \begin{pmatrix} 1 & 0 & 0 \\ f & 1 & 0 \\ 0 & 0 & 1 \end{pmatrix} \begin{pmatrix} 1 & 0 & 0 \\ 0 & a^{-1} & 0 \\ -c^{-1}e & -c^{-1}ba^{-1} & c^{-1} \end{pmatrix} \theta \\ &= \begin{pmatrix} 1 & 0 & 0 \\ Af & Aa^{-1} & 0 \\ Bf - Cc^{-1}e & Ba^{-1} - Cc^{-1}ba^{-1} & Cc^{-1} \end{pmatrix} \theta = m\theta. \end{aligned}$$

Hence we may normalize all of the entries of m by choosing

$$a = A, \qquad c = C, \qquad b = B, \qquad e = Bf$$

and since $f \neq 0$,

$$Af = {}^t(1, 0, \cdots, 0).$$

This yields the subgroup G_1 and coframe given by

$$\hat{\omega} = \begin{pmatrix} 1 & 0 & 0 \\ 0 & A & 0 \\ 0 & B & C \end{pmatrix} \begin{pmatrix} dt \\ a_0\, dx \\ du \end{pmatrix},$$

where

$$a_0 f = {}^t(1, 0, \cdots, 0)$$

and A has the form

$$A = \begin{pmatrix} 1 & b_1 \\ 0 & A_1 \end{pmatrix}.$$

As the 1-forms dt and dT are invariant and decouple from the rest of the problem, it suffices to consider the conditions in terms of the remaining components of the coframes. Note, however, it was absolutely essential to keep dt and dT until the reduction to a determined problem. Thus it suffices to consider the lifted problem

$$\begin{pmatrix} \eta \\ \mu \end{pmatrix} = \begin{pmatrix} 1 & b_1 & 0 \\ 0 & A_1 & 0 \\ b & b_2 & A_2 \end{pmatrix} \begin{pmatrix} a_0\, dx \\ du \end{pmatrix}.$$

The notation was not randomly chosen; if we change the order of the first row and the next $(m-1)$ rows we find

$$\begin{pmatrix} A_1 & 0 & 0 \\ b_1 & 1 & 0 \\ b & b_2 & A_2 \end{pmatrix}.$$

We have already seen a group with this block decomposition in Lectures 1 and 5. This is the group that appeared in the particle Lagrangian problem. Thus if $\Gamma(m,p)$ is the G associated to a control problem for m states and p controls and $G(m)$ is the G associated to the Lagrangian particle problem for curves in $\mathbf{R}^m$, we have

$$\Gamma(n, n-1) \simeq G(n-1).$$

The equivalence problem then implies there is a possible geometric isomorphism

$$\begin{pmatrix} \text{Geometry of control systems} \\ \text{with } m \text{ states and } m-1 \text{ controls} \\ \text{under feedback} \end{pmatrix} \simeq \begin{pmatrix} \text{Geometry of Lagrangian} \\ \text{particle theories} \\ \end{pmatrix}$$

and this accounts for the strange title of this lecture. The existence of such geometric isomorphisms was anticipated by Felix Klein, and made possible by É. Cartan. The following citation is most appropriate at this time:

> But the great originality of Klein is to have conceived the relation between a "geometry" and its group by reversing the roles of these two entities, the group becoming then the fundamental object, and the various spaces on which it "operates" showing various aspects of the structure of the group: an idea whose fecundity he already demonstrated by establishing the "isomorphism" of "geometries" of entirely different appearance, for example, the conformal geometry of the space of three dimensions and the hyperbolic non-euclidean geometry of four dimensions.[3] In fact, this principle had a significance much greater than Klein could have imagined, as subsequent developments were to reveal.[4]

A closer look at the two equivalence problems shows that there is a rank condition hidden in the local coframes which obstructs the correspondence

---

[3]See É. Cartan, *Lessons On Complex Projective Geometry*, Gauthier-Villars, 1931, and J. Dieudonné, *Linear Algebra and Elementary Geometry*, Annexe III, Hermann, 1968.

[4]Introduction, by Jean Dieudonné, to *The Erlangen Program* by Felix Klein, translation by my REU assistant Adam Falk.

of some control problems with m states with Lagrangian particle problems. We will uncover this condition after we do part of the equivalence problem.

In order to be concrete, reorder the equations so that $f_1 \neq 0$ and choose

$$a_0 = \begin{pmatrix} 1/f_1 & 0 & \cdots & 0 \\ -f_2 & f_1 & \cdots & 0 \\ \vdots & \vdots & \ddots & \vdots \\ -f_m & 0 & \cdots & f_1 \end{pmatrix}$$

with zeros off the diagonal and first column. Then

$$a_0 f = {}^t(1, 0, \cdots, 0).$$

Now with $G = \Gamma(m, p)$

$$dSS^{-1} = \begin{pmatrix} 0 & * & 0 \\ 0 & * & 0 \\ * & * & * \end{pmatrix}.$$

Let

$$\eta = \begin{pmatrix} \eta^1 \\ \vec{\eta} \end{pmatrix};$$

then, using the principal components, we have

$$d\begin{pmatrix} \eta^1 \\ \vec{\eta} \\ \mu \end{pmatrix} = \begin{pmatrix} 0 & \vec{\alpha}^1 & 0 \\ 0 & \hat{\alpha} & 0 \\ \beta_1 & \beta & \gamma \end{pmatrix} \wedge \begin{pmatrix} \eta^1 \\ \vec{\eta} \\ \mu \end{pmatrix} + {}^t\mu \begin{pmatrix} \overrightarrow{M}^1 \\ \widehat{M} \\ 0 \end{pmatrix} \eta + {}^t\eta \begin{pmatrix} \overrightarrow{L}^1 \\ \hat{L} \\ 0 \end{pmatrix} \eta + {}^t\mu \begin{pmatrix} \overrightarrow{N}^1 \\ \widehat{N} \\ 0 \end{pmatrix}.$$

There are integrability conditions given by

$$d\eta = d(Aa_0\, dx) = (dAA^{-1} + A\, da_0 a_0^{-1} A^{-1}) \wedge \eta$$

which imply there are no quadratic terms in μ and $\overrightarrow{N} = 0$ and $\widehat{N} = 0$.

The terms

$${}^t\eta \begin{pmatrix} \overrightarrow{L} \\ \hat{L} \\ 0 \end{pmatrix} \eta$$

can be absorbed in α since $\eta^1 \wedge \eta^1$ is the only term not visibly absorbable, and it is zero. Similarly, all the terms

$${}^t\mu \begin{pmatrix} \overrightarrow{M}^1 \\ \widehat{M} \\ 0 \end{pmatrix} \eta$$

can be absorbed except

$${}^t\mu M \eta^1.$$

The infinitesimal actions on $\vec{M}^1$ and $\widehat{M}$ are

$$\left.\begin{aligned} d\vec{M}^1 - \vec{\alpha}^1\widehat{M} + \vec{M}^1\gamma \equiv 0 \\ d\widehat{M} - \hat{\alpha}\widehat{M} + \widehat{M}\gamma \equiv 0 \end{aligned}\right. \quad \text{mod base.}$$

If we compute parametrically

$$M = -AA_0\frac{\partial(f_1,\cdots,f_m)}{\partial(u^1,\cdots,u^p)}A_2^{-1},$$

hence the rank of M is equal to the rank of the partial Jacobian. Now let us focus in on the systems with m states and $m-1$ controls which are geometrically isomorphic to the Lagrangian particle problem. Thus we assume rank $M = p$ and pursue the case $p = m - 1$. Since

$$M = \begin{pmatrix} \vec{M}^1 \\ \widehat{M} \end{pmatrix}$$

there are two possibilities as follows.

Case I. rank $\widehat{M} = p$

or

Case II. rank $\widehat{M} = p - 1$.

In the Lagrangian particle problem we saw that the rank of the matrix $\widehat{M}$ was always maximal rank, and hence Case II control systems do not give rise to classical Lagrangians. They do, however, correspond to variational problems with nonholonomic constraints as discussed in [Gr. 1983].

If the control problem actually depends on all p-controls then rank $M = p$. The problem for $p = m$ is uninteresting since a direct argument shows that

$$\frac{dx}{dt} = u$$

is a normal form.

If rank $\widehat{M} = p$ then $\vec{\alpha}^1\widehat{M}$ contains p independent infinitesimal translations, and we can normalize

$$\vec{M}^1 = 0.$$

The action on $\widehat{M}$ is by infinitesimal left multiplication by an $m \times m$ matrix and infinitesimal right multiplication by a $p \times p$ matrix. Hence we can normalize

$$\widehat{M} = \begin{pmatrix} 0 \\ I_p \end{pmatrix},$$

where I_p is the $p \times p$ identity. The structure equations then become

$$d\begin{pmatrix} \eta^1 \\ \vec{\eta} \\ \mu \end{pmatrix} = \begin{pmatrix} 0 & \vec{\alpha}^1 & 0 \\ 0 & \hat{\alpha} & 0 \\ \beta_1 & \beta & \gamma \end{pmatrix} \wedge \begin{pmatrix} \eta^1 \\ \vec{\eta} \\ \mu \end{pmatrix} + \begin{pmatrix} 0 \\ \eta^1 \wedge \mu \\ 0 \end{pmatrix}.$$

These normalizations induce new principal components

$$\gamma - \hat{\alpha},\ \vec{\alpha}^1.$$

If we differentiate mod $\vec{\eta}$ we find integrability conditions

$$\vec{\alpha}^1 \wedge \mu \equiv 0 \quad \operatorname{mod}(\eta^1, \vec{\eta}).$$

Hence if we let

$$\vec{\alpha}^1 = a\,\eta^1 + {}^t\vec{\eta}B + {}^t\mu C,$$

we know by Cartan's lemma that ${}^tC = C$ and

$$d\begin{pmatrix} \eta^1 \\ \vec{\eta} \\ \mu \end{pmatrix} = \begin{pmatrix} 0 & 0 & 0 \\ 0 & \hat{\alpha} & 0 \\ \beta_1 & \beta & \gamma \end{pmatrix} \wedge \begin{pmatrix} \eta^1 \\ \vec{\eta} \\ \mu \end{pmatrix} + \begin{pmatrix} a\eta^1\vec{\eta} + {}^t\vec{\eta}B\vec{\eta} + {}^t\mu C\vec{\eta} \\ \eta^1 \wedge \mu \\ 0 \end{pmatrix}.$$

Note that we see the form η^1 is now invariant. Differentiating as usual gives

$$\begin{aligned} &da + a\hat{\alpha} + {}^t\beta_1 C \equiv 0 \\ &dB + {}^t\hat{\alpha}B - B\hat{\alpha} - \tfrac{1}{2}({}^t\beta C - C\beta) \equiv 0 \quad \text{mod base.} \\ &dC - {}^t\hat{\alpha}C - C\hat{\alpha} \equiv 0 \end{aligned}$$

The generic case occurs when $\det C \neq 0$. This corresponds to the regular problem in the Lagrangian problem. In this case we can normalize C by the third set of infinitesimal actions to $C = Q$, where Q is a diagonal matrix with ones or minus ones on the diagonal. The first equation then has an infinitesimal translation so that we can normalize $a = 0$, and since B can be taken skew symmetric, the terms ${}^t\beta C - C\beta$ contain enough translations to normalize $B = 0$.

In order to make the exposition simpler let us assume $Q = \mathbf{I}$, then

$$d\eta^1 = {}^t\mu \wedge \vec{\eta}.$$

This is enough information to get a dramatic result from the geometric isomorphism. It is clear that the 1-form η^1 corresponds to $L\,dx$, and hence it is natural to ask what is the corresponding variational problem. Since

$$\eta = AA_0 dx$$

we have

$$AA_0(dx - f\,dt) = \eta - \begin{pmatrix} 1 \\ 0 \\ \vdots \\ 0 \end{pmatrix} dt$$

and the system

$$\{dx - f\,dt\} = \{\eta^1 - dt, \eta^2, \cdots, \eta^m\},$$

as such along integral curves of the system

$$\eta^1 = dt,$$

and

$$\int_{\substack{integral\\curve}} \eta^1 = \text{time along integral curve.}$$

Since $d\eta^1 = {}^t\mu \wedge \vec{\eta}$ it is natural to index $\vec{\eta} = {}^t(\eta^2, \cdots, \eta^m)$ and $\mu = {}^t(\mu^2, \cdots, \mu^m)$ so that

$$d\eta^1 = \sum \mu^\alpha \wedge \eta^\alpha, \qquad 2 \le \alpha \le m.$$

The system $\{\mu^2, \cdots, \mu^m, \eta^2, \cdots, \eta^m\}$ is completely integrable and gives the Euler–Lagrange equations for the variational problem (see [G. 1983])

$$\int_{\substack{all\\curves}} \eta^1.$$

Now

$$\mu \equiv T\, dg \quad \operatorname{mod} \vec{\eta},$$

where $g = {}^t(g_2, \cdots, g_m)$ with $g_\alpha = g_\alpha(x, u)$ and

$$0 \ne \mu^2 \wedge \cdots \wedge \mu^m \wedge \eta^1 \wedge \cdots \wedge \eta^m = \det T \det \frac{\partial g}{\partial u} du^2 \wedge \cdots \wedge du^m \wedge \eta^1 \wedge \cdots \wedge \eta^m$$

hence $\det \partial g/\partial u \ne 0$, and we can solve

$$g = c, \qquad c, \text{ a constant vector}$$

for $u = u(x, c)$. We have seen the solutions of

$$\frac{dx}{dt} = f(x, u(x, c))$$

necessarily solve

$$\vec{\eta} = 0$$

and by construction also solve

$$\mu = 0.$$

Therefore the solutions are time critical among all curves and in particular time critical among all integral curves. The function $u = u(x, c)$ is called a *time-critical closed loop control*, which is important for engineering applications.

Now let us return to the equivalence problem, which in the alter ego of the Lagrangian problem is joint work with R. Bryant. After the second-order normalizations just described, we have the first-order principal components

$$\gamma - \hat{\alpha},$$

and the new second-order principal components

$${}^t\beta_1, \quad {}^t\beta - \beta, \quad {}^t\hat{\alpha} + \hat{\alpha}.$$

After some significant algebra to *arrange natural absorptions*, we have structure equations

$$d\begin{pmatrix} \eta^1 \\ \vec{\eta} \\ \mu \end{pmatrix} = \begin{pmatrix} 0 & 0 & 0 \\ 0 & \hat{\alpha} & 0 \\ 0 & \beta & \hat{\alpha} \end{pmatrix} \wedge \begin{pmatrix} \eta^1 \\ \vec{\eta} \\ \mu \end{pmatrix} + \begin{pmatrix} {}^t\mu \wedge \vec{\eta} \\ \eta^1 \wedge \mu \\ M\,\eta^1 \wedge \mu \end{pmatrix}$$

with refined congruences

$$\begin{aligned} {}^t\hat{\alpha} + \hat{\alpha} &\equiv M\,\eta^1 \mod \vec{\eta} \\ {}^tM &= M. \end{aligned}$$

The infinitesimal action on the torsion is

$$dM + M\,\hat{\alpha} - \hat{\alpha}\,M - (\beta + {}^t\beta) \equiv 0 \mod \text{base}$$

and hence we can translate M to zero:

$$M = 0.$$

This introduces new principal components β and the refined congruence

$${}^t\hat{\alpha} + \hat{\alpha} \equiv 0 \mod \vec{\eta}.$$

Finally after more significant algebra we have the structure equations in the form

$$d\begin{pmatrix} \eta^1 \\ \vec{\eta} \\ \mu \end{pmatrix} \wedge \begin{pmatrix} \eta^1 \\ \vec{\eta} \\ \mu \end{pmatrix} = \begin{pmatrix} 0 & 0 & 0 \\ 0 & \phi & 0 \\ 0 & 0 & \phi \end{pmatrix} + \begin{pmatrix} {}^t\mu \wedge \vec{\eta} \\ \eta^1 \wedge \mu + {}^t\lambda \wedge \mu \\ {}^t\zeta \wedge \eta \end{pmatrix},$$

where

(1) ϕ is now unique, hence there is an e-structure,
(2) $(\lambda)_{ab} = \sum S_{abc}\eta^c$ with S_{abc} totally symmetric,
(3) $(\zeta)_{ab} = B_{abc}\eta^c + \sigma_{ab}\eta^1 + T_{abc}\mu^c$

with

$$\sigma_{ab} = \sigma_{ba}, \quad T_{abc} = T_{bca}, \quad B_{abc} = -B_{bac}, \quad \text{and} \quad B_{abc} + B_{bca} + B_{cab} = 0.$$

Therefore the fundamental invariants of the problem are

$$\begin{aligned} &\sigma = (\sigma_{ab}) && \text{a symmetric 2-tensor} \\ &S = (S_{abc}) && \text{a symmetric 3-tensor} \\ &\left.\begin{aligned} T &= (T_{abc}) \\ B &= (B_{abc}) \end{aligned}\right\} && \text{two 3-tensors with some symmetries.} \end{aligned}$$

The control theoretical interpretation of these invariants has not yet been worked out.

LECTURE 9

Normal Forms and Generalized Geometries

To get a better feel for the results uncovered in the last section we now specialize to the simplest cases and indicate the type of results toward which the program is aiming.

Let us consider the equivalence problem for two states and one control which is joint work with W. Shadwick [G–S. 1987]. This is the case $m = 2$, $p = 1$, and is the simplest case where $p = m - 1$. Since there are interesting results in the degenerate cases, let us go over the flow of the ideas, and give the results of the parametric calculations.

The initial coframe is

$$\begin{pmatrix} \eta_U^1 \\ \eta_U^2 \\ \mu_U \end{pmatrix} = \begin{pmatrix} A_0\, dx \\ du \end{pmatrix} = \begin{pmatrix} \frac{1}{f^1} & 0 & 0 \\ -f^2 & f^1 & 0 \\ 0 & 0 & 1 \end{pmatrix} \begin{pmatrix} dx^1 \\ dx^2 \\ du \end{pmatrix} = \begin{pmatrix} \frac{1}{f^1}\, dx^1 \\ -f^2\, dx^1 + f^1\, dx^2 \\ du \end{pmatrix}$$

and the lifted coframe is

$$\begin{pmatrix} \eta^1 \\ \eta^2 \\ \mu \end{pmatrix} = \begin{pmatrix} 1 & b_1 & 0 \\ 0 & A_1 & 0 \\ b & b_2 & A_2 \end{pmatrix} \begin{pmatrix} \eta_U^1 \\ \eta_U^2 \\ \mu_U \end{pmatrix}.$$

The first structure equation is

$$d \begin{pmatrix} \eta^1 \\ \eta^2 \\ \mu \end{pmatrix} = \begin{pmatrix} 0 & \alpha_1 & 0 \\ 0 & \alpha_2 & 0 \\ \beta_1 & \beta & \gamma \end{pmatrix} \wedge \begin{pmatrix} \eta^1 \\ \eta^2 \\ \mu \end{pmatrix} + \begin{pmatrix} m^1 \\ m^2 \\ 0 \end{pmatrix} \eta^1 \wedge \mu.$$

If we let

$$\Delta = f^2 f^1_{,u} - f^1 f^2_{,u}$$

then

$$\begin{pmatrix} m^1 \\ m^2 \end{pmatrix} = \frac{1}{A_2} \begin{pmatrix} -(\ln|f^1|)_u + A_1 \Delta \\ A_1\, \Delta \end{pmatrix}.$$

The infinitesimal action on torsion is

$$dm^1 - \alpha_1\, m^2 + \gamma\, m^1 \equiv 0$$
$$dm^2 - (\alpha_2 - \gamma) m^2 \equiv 0.$$

There are several degenerate cases that appear.

Case I. $m^1 = 0$, $m^2 = 0$.

There is no torsion, hence we test for involution. The reduced tableau matrix is

$$\begin{pmatrix} 0 & \alpha_1 & 0 \\ 0 & \alpha_2 & 0 \\ \beta_1 & \beta & \gamma \end{pmatrix}$$

and the reduced Cartan characters are

$$\sigma_1 = 3, \qquad \sigma_2 = 1, \qquad \sigma_3 = 1$$

and g is

$$\sigma = 3 + 2 \cdot 1 + 3 \cdot 1 = 8.$$

The degree of indeterminancy in $\alpha_1 = 1$ and in $\alpha_2 = 1$, and by Cartan's lemma the degree of indeterminancy in β_1, β, $\gamma = 6$. Hence $\dim \mathcal{G}^{(1)} = 8$, and the system is in involution. A normal form for this class is

$$\frac{dx^1}{dt} = 1, \qquad \frac{dx^2}{dt} = 0.$$

Case II. $m^1 \neq 0$, $m^2 = 0$.

Then we can normalize m^1 to 1 which induces the new principal component, γ.

The reduced tableau matrix is

$$\begin{pmatrix} 0 & \alpha_1 & 0 \\ 0 & \alpha_2 & 0 \\ \beta_1 & \beta & 0 \end{pmatrix}$$

and

$$\sigma_1 = 3, \qquad \sigma_2 = 1,$$

so σ is $\sigma = 5$. Visibly $\dim \mathcal{G}^{(1)} = 5$ and this system is again in involution. A normal form for this class is

$$\frac{dx^1}{dt} = u, \qquad \frac{dx^2}{dt} = 0.$$

Now assume $m^2 \neq 0$, then we can normalize $m^2 = 1$ and $m^1 = 0$ which induces new principal components

$$\gamma - \alpha_2, \qquad \alpha_1.$$

The normalizations force relations on the group G in the form

$$A_2 = A_1 \Delta, \qquad b_1 = \frac{f^1_{,u}}{f^1 \Delta}.$$

We choose an adapted coframe by choosing a coframe of the form

$$\begin{pmatrix} 1 & \frac{f^1_{,u}}{f^1\Delta} & 0 \\ 0 & A_1\Delta & 0 \\ b & b_2 & A_1 \end{pmatrix} \begin{pmatrix} \eta^1_U \\ \eta^2_U \\ \mu_U \end{pmatrix}.$$

If we choose $A_1 = 1$, $b = 0$, $b_2 = 0$, we get

$$\begin{pmatrix} \frac{f^1_{,u}}{\Delta}\,dx^2 - \frac{f^2_{,u}}{\Delta}\,dx^1 \\ (f^1\,dx^2 - f^2\,dx^1) \\ \Delta\,du \end{pmatrix}$$

and this gives rise to the structure equations

$$d\begin{pmatrix} \eta^1 \\ \eta^2 \\ \mu \end{pmatrix} = \begin{pmatrix} 0 & 0 & 0 \\ 0 & \alpha_2 & 0 \\ \beta_1 & \beta & \alpha_2 \end{pmatrix} \wedge \begin{pmatrix} \eta^1 \\ \eta^2 \\ \mu \end{pmatrix} + \begin{pmatrix} 0 \\ 1 \\ 0 \end{pmatrix} \eta^1\wedge\mu + \begin{pmatrix} a \\ 0 \\ 0 \end{pmatrix} \eta^1\wedge\eta^2 + \begin{pmatrix} c \\ 0 \\ 0 \end{pmatrix} \mu\wedge\eta^2.$$

These new torsion terms can be calculated parametrically and are

$$a = \frac{1}{A_1}\left(\left(\frac{f^1_{,u}}{\Delta}\right)_{x^1} + \left(\frac{f^2_{,u}}{\Delta}\right)_{x^2} - \frac{cb}{f^1}\right)$$
$$c = \frac{1}{\Delta^3 A_2^2}\left(f^2_{,u} f^1_{,u,u} - f^1_{,u} f^2_{,u,u}\right).$$

These undergo infinitesimal fiber action

$$\begin{aligned} da + a\,\alpha_2 + c\,\beta_1 &\equiv 0 \\ dc + 2c\,\alpha_2 &\equiv 0 \end{aligned} \qquad \text{mod base.}$$

Case III. $a = 0$, $c = 0$.

This means η^1 is invariant and holds if and only if $d\eta^1 = 0$ and hence η^1 is a conservation law for the problem. The reduced tableau matrix is

$$\begin{pmatrix} 0 & 0 & 0 \\ 0 & \alpha_2 & 0 \\ \beta_1 & \beta & \alpha_2 \end{pmatrix}$$

and

$$\sigma_1 = 2, \qquad \sigma_2 = 1$$

with g

$$2 + 3 \cdot 1 = 5.$$

Visibly $\dim \mathcal{G}^{(1)} = 5$ and the system is in involution. A normal form for this class is

$$\frac{dx^1}{dt} = u, \qquad \frac{dx^2}{dt} = x^2.$$

Case IV. $a \neq 0$, $c = 0$.

Then we can normalize a to one which induces a new principal component, α_2. This normalization forces relations on the group of the form

$$A_1 = \left(\frac{f^1_{,u}}{\Delta}\right)_{x^1} + \left(\frac{f^2_{,u}}{\Delta}\right)_{x^2}.$$

We choose an adapted coframe of the form

$$\begin{pmatrix} \frac{f^1_{,u}}{\Delta}dx^2 - \frac{f^2_{,u}}{\Delta}dx^1 \\ A_1\Delta(f^1dx^2 - f^2dx^1) \\ A_1\Delta\, du \end{pmatrix}.$$

This gives rise to structure equations

$$d\begin{pmatrix}\eta^1\\ \eta^2\\ \mu\end{pmatrix} = \begin{pmatrix}0&0&0\\0&0&0\\ \beta_1&\beta&0\end{pmatrix}\wedge\begin{pmatrix}\eta^1\\ \eta^2\\ \mu\end{pmatrix} + \begin{pmatrix}1\\ g\\ 0\end{pmatrix}\eta^1\wedge\eta^2 + \begin{pmatrix}0\\1\\0\end{pmatrix}\mu\wedge\eta^1.$$

Note there is no $\mu\wedge\eta^2$ torsion because of the integrability condition $d^2 = 0$.

The torsion term g has infinitesimal fiber action

$$dg - \beta \equiv 0 \quad \text{mod base},$$

and hence we can translate g to zero, which introduces new torsion

$$d\begin{pmatrix}\eta^1\\ \eta^2\\ \mu\end{pmatrix} = \begin{pmatrix}0&0&0\\0&0&0\\ \beta_1&0&0\end{pmatrix}\wedge\begin{pmatrix}\eta^1\\ \eta^2\\ \mu\end{pmatrix} + \begin{pmatrix}1\\0\\0\end{pmatrix}\eta^1\wedge\eta^2 + \begin{pmatrix}0\\1\\0\end{pmatrix}\mu\wedge\eta^1 + \begin{pmatrix}0\\0\\s\end{pmatrix}\mu\wedge\eta^2.$$

There is, however, an integrability condition which forces $s = -1$. Notice this class is characterized by $d\eta^1\wedge\eta^1 = 0$. The structure equations are visibly in involution and a normal form for this class is

$$\frac{dx^1}{dt} = u, \qquad \frac{dx^2}{dt} = x^1$$

which is the controllable Brunovski normal form. Note that $c = 0$ if and only if

$$f^2_{,u}f^1_{,u,u} - f^1_{,u}f^2_{,u,u} = 0,$$

and hence all affine linear systems

$$\frac{dx}{dt} = Ax + bu$$

are in Cases I–IV and as a result are equivalent to linear systems. In particular, no equation equivalent to affine linear can satisfy $c \neq 0$.

These linear normal forms were known to control theorists, but in the setting of a control linear feedback which involves quite different methods. It is also possible to use C^∞ normal forms in exterior differential systems to get these results. The Cartan–Kähler theorem was used since it is the general technique.

Case V. $c \neq 0$.

In this case we can normalize c to plus or minus one and then translate a to zero. Thus

$$c = \varepsilon f^1 \left(\left(\frac{f^1_{,u}}{\Delta} \right)_{x^1} + \left(\frac{f^2_{,u}}{\Delta} \right)_{x^2} \right)$$

$$A^2_1 = \left| \frac{f^2_{,u} f^1_{,u,u} - f^1_{,u} f^2_{,u,u}}{\Delta^3} \right|,$$

where

$$\varepsilon = \operatorname{sign} \left(\frac{f^2_{,u} f^1_{,u,u} - f^1_{,u} f^2_{,u,u}}{\Delta^3} \right).$$

This introduces new torsion

$$d \begin{pmatrix} \eta^1 \\ \eta^2 \\ \mu \end{pmatrix} = \begin{pmatrix} 0 & 0 & 0 \\ 0 & 0 & 0 \\ 0 & \beta & 0 \end{pmatrix} \wedge \begin{pmatrix} \eta^1 \\ \eta^2 \\ \mu \end{pmatrix} + \begin{pmatrix} 0 \\ l \\ 0 \end{pmatrix} \eta^1 \wedge \eta^2 + \begin{pmatrix} 0 \\ 1 \\ n \end{pmatrix} \mu \wedge \eta^1 + \begin{pmatrix} \varepsilon \\ I \\ 0 \end{pmatrix} \mu \wedge \eta^2$$

and again there is an integrability condition $d^2 = 0$ which forces $l = n$.

The infinitesimal fiber action only appears on the coefficient n and is

$$dn - \beta \equiv 0 \quad \text{mod base.}$$

Normalizing n to zero by translation results in the principal component, β, and results in an e-structure. The final structure equations are

$$\begin{aligned} d\eta^1 &= \varepsilon \mu \wedge \eta^2 \\ d\eta^2 &= \mu \wedge \eta^1 + I\, \mu \wedge \eta^2 \\ d\mu &= -K \eta^1 \wedge \eta^2 + J\, \mu \wedge \eta^2. \end{aligned}$$

If we define generalized derivatives

$$dh = h_{\eta^1}\, \eta^1 + h_{\eta^2}\, \eta^2 + h_\mu\, \mu$$

we see there are integrability conditions

$$J = -I_{\eta^1}, \qquad J_{\eta^1} = -K_\mu - KI.$$

The cases when all the invariants are constant are the cases when there is a transitive Lie group of equivalences. There are two cases.

Case A. $I = 0$, whence $J = 0$.

$$\begin{aligned} d\eta^1 &= \varepsilon\,\mu \wedge \eta^2 \\ d\eta^2 &= \mu \wedge \eta^1 \\ d\mu &= -K\,\eta^1 \wedge \eta^2. \end{aligned}$$

Here is another unexpected geometric isomorphism, because those are the structure equations of the pseudo–Riemannian metric

$$ds^2 = (\eta^1)^2 \pm \epsilon(\eta^2)^2$$

with Gauss curvature K. Note that the integrability condition gives $K_\mu = 0$, and hence K is a function on (η^1, η^2)–space.

Case B. $I = \kappa = \text{constant} \neq 0$, whence $J = 0$ and $K = 0$.

$$\begin{aligned} d\eta^1 &= \varepsilon\,\mu \wedge \eta^2 \\ d\eta^2 &= \mu\eta^1 + \kappa\,\mu \wedge \eta^2 \\ d\mu &= 0. \end{aligned}$$

This is an example of a "generalized geometry" with mixed torsion. The term $\kappa\,\mu \wedge \eta^2$ is vertizontal and not quadratic in $\eta^1 \wedge \eta^2$, as would be the torsion of a metric.

In the general case we have

$$\begin{aligned} d\eta^1 &= \pm\mu \wedge \eta^2 \\ d\eta^2 &= \mu \wedge \eta^1 + I\,\mu \wedge \eta^2 \\ d\mu &= -K\,\eta^1 \wedge \eta^2 + J\,\mu \wedge \eta^2. \end{aligned}$$

These are the structure equations of a "generalized geometry" with mixed torsion and mixed curvature. Again the terms $I\,\mu \wedge \eta^2$ and $J\,\mu \wedge \eta^2$ are vertizontal.

This structure was known to Cartan in the alter ego of Lagrangian particle theory in the plane. His paper [C. 1930] considers the geometric meaning of this generalized geometry, but this needs work to put it on a modern foundation similar to the understanding of connections.

In summary, we have four cases of systems in involution and one case of an identity structure. We summarize this in diagram form (see Fig. 3).

There are also normal form questions that are natural to ask about Case V. In particular, the control systems of this type that have the simplest normal forms would be those with the maximal Lie symmetry.

As we have seen, there are only two such cases.

Restricting to $\varepsilon = +1$ normal forms are given by the following.

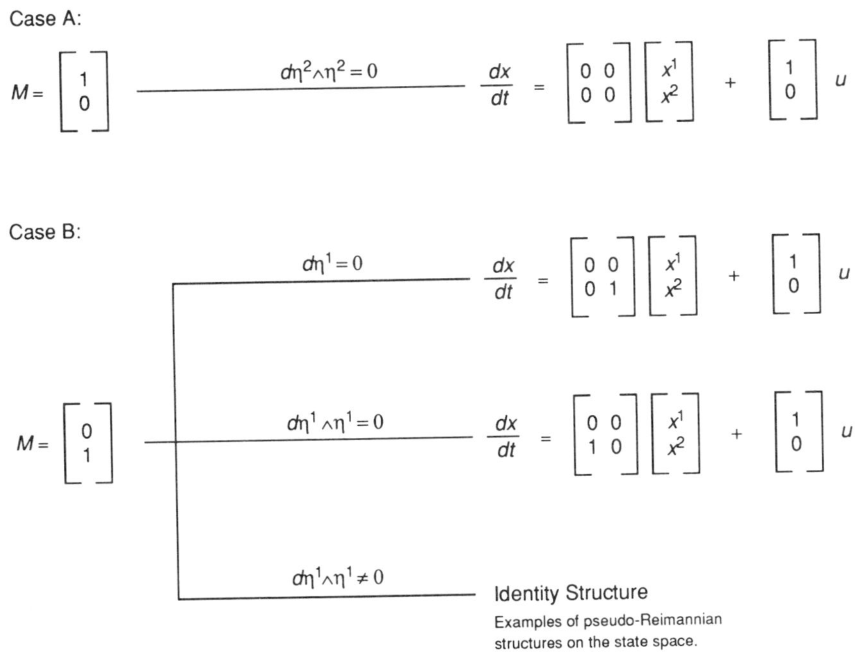

FIG. 3. *Schematic of 2-state 1-input control systems.*

Case A. $I = 0$, $J = 0$, $K =$ constant.

$$\frac{dx^1}{dt} = \frac{\sin u}{1 - K/4((x^1)^2 + (x^2)^2)}$$
$$\frac{dx^2}{dt} = \frac{\cos u}{1 - K/4((x^1)^2 + (x^2)^2)}.$$

The case $\varepsilon = -1$ is obtained by adding h to sin and cos.

Case B. $J = 0$, $K = 0$, $I =$ constant.

$$\frac{dx^1}{dt} = \sinh(\alpha u)e^{\beta u} \qquad I = -\frac{\beta}{\sqrt{\varepsilon(\beta^2 - \alpha^2)}} \qquad \alpha \neq \beta.$$
$$\frac{dx^2}{dt} = \cosh(\alpha u)e^{\beta u}$$

The next case of control systems having an alter ego in Lagrangian particle theory is the case of three states and two controls. The initial torsion invariant M has one of two normal forms

$$M = \begin{pmatrix} 1 & 0 \\ 0 & 1 \\ 0 & 0 \end{pmatrix} \quad \text{or} \quad \begin{pmatrix} 0 & 0 \\ 1 & 0 \\ 0 & 1 \end{pmatrix}$$

and the lifted coframes consist of 1-forms

$$\eta^1, \eta^2, \eta^3 \quad \text{and} \quad \mu^1, \mu^2.$$

The abstract equivalence problem has been solved in George Wilkens' dissertation [Wi. 1987] and includes ferreting out some beautiful geometric structures. His work is outlined in diagram form (see Fig. 4).

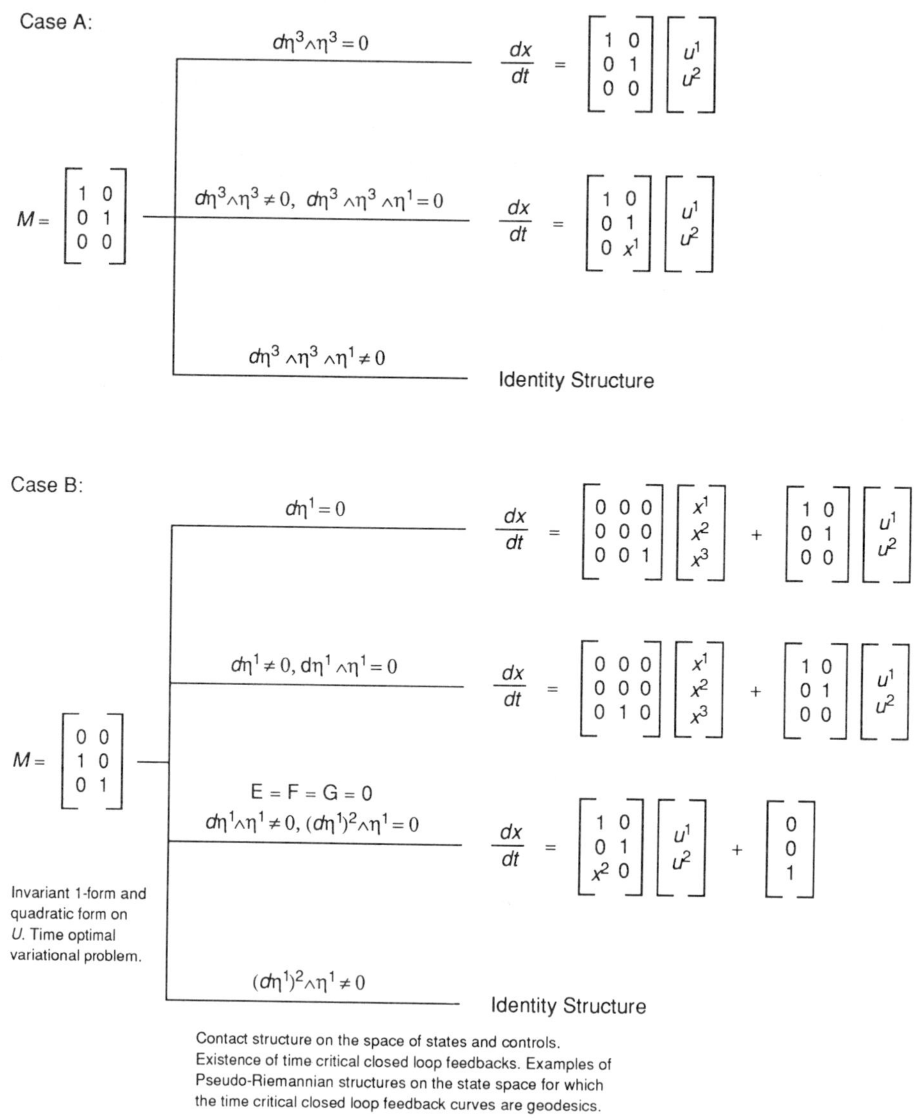

FIG. 4. *Schematic of 3-state, 2-input control systems (Wilkens' dissertation).*

Note that two of the normal forms are bilinear, and all of the linear systems are characterized by

$$d\eta^3 \wedge \eta^3 = 0 \quad \text{or} \quad d\eta^1 \wedge \eta^1 = 0,$$

hence there are examples of bilinear systems that are not equivalent to linear systems under full nonlinear feedback.

LECTURE 10

Prolongation

The final piece of the puzzle is the procedure to follow if the normalizations and reductions are completed, group parameters remain, but the system is not in involution. The idea is simple: we have a coframe on $U \times G$ defined up to a group $G^{(1)}$, hence lift the problem to $U \times G \times G^{(1)}$ and start the whole procedure afresh.

Since there is nothing in this procedure that we have not discussed, let us immediately look at some examples.

Example 2. *Conformal geometry* (*continued*). The Maurer–Cartan matrix,

$$\varpi(S) = \frac{d\lambda}{\lambda}\mathbf{I} + dSS^{-1},$$

is characterized by the defining relation

$$\varpi + {}^t\varpi - \frac{2}{n}\mathrm{tr}\varpi\,\mathbf{I} = 0.$$

If we have

$$d\omega = \Delta \wedge \omega$$

on $U \times G$, then the principal components of order 1 are

$$\Delta + {}^t\Delta - \frac{2}{n}\mathrm{tr}\Delta\,\mathbf{I}.$$

As such

$$d\omega = \delta \wedge \omega + \psi \wedge \omega,$$

where

$$\delta = \frac{1}{2}(\Delta - {}^t\Delta) + \frac{2}{n}\mathrm{tr}\Delta\,\mathbf{I} \quad \text{and} \quad \psi = \frac{1}{2}(\Delta + {}^t\Delta) - \frac{2}{n}\mathrm{tr}\Delta\,\mathbf{I}.$$

The $SO(p,q)$ lemma implies there is a unique solution Π to the exterior equations

$$\Pi \wedge \omega = \psi \wedge \omega$$
$$\Pi + {}^t\Pi = 0.$$

Let $\phi = \frac{1}{2}(\Delta - {}^t\Delta) + \Pi$ and $\pi = \frac{2}{n}\text{tr}\Delta$ then

$$d\omega = (\pi\mathbf{I} + \phi) \wedge \omega \quad \text{with} \quad {}^t\phi = -\phi$$

and all torsion is absorbed. Now we need to check for involution. The reduced Cartan characters are

$$\sigma_1 = n, \qquad \sigma_2 = n-2, \cdots, \sigma_{n-1} = 1$$

and

$$\sigma = \sum_{j=1}^{n-1} j\sigma_j = n + \sum_{j=2}^{n-1} j(n-j).$$

The group $G^{(1)}$ gives the ambiguity in the forms π and ϕ. Thus let

$$d\omega = (\pi'\mathbf{I} + \phi') \wedge \omega \quad \text{with} \quad {}^t\phi' = -\phi'.$$

Subtraction gives

$$0 = ((\pi' - \pi)\mathbf{I} + \phi' - \phi) \wedge \omega$$

and wedging with $(n-1)$ omegas gives

$$0 = (\pi' - \pi) \wedge \omega^1 \wedge \cdots \wedge \omega^n.$$

This means there is a vector a satisfying

$$\pi' - \pi = a\omega.$$

Substituting this back into the subtracted equation gives

$$\begin{aligned} 0 &= (a\omega\mathbf{I} + \phi' - \phi) \wedge \omega \\ &= ({}^ta{}^t\omega - \omega a + \phi' - \phi) \wedge \omega. \end{aligned}$$

Exercise. Check that if $a = \{a_i\}$ then this last equation becomes

$$\sum (a_i\omega^j - a_j\omega^i + \phi'^i_j - \phi^i_j) \wedge \omega^j = 0.$$

Since the expression inside the parentheses is now skew symmetric, the S_3 lemma gives

$$\begin{aligned} \phi' - \phi &= \omega a - {}^ta{}^t\omega \\ &= A(a)\omega. \end{aligned}$$

Exercise. Let $a = \{a_i\}$ and $A(a) = \{A_i{}^j{}_k\}$, then show

$$A_i{}^j{}_k = a_i\delta^j_k - a_j\delta^i_k.$$

Collecting these results we have

$$\begin{pmatrix} \pi' \\ \phi' \\ \omega \end{pmatrix} = \begin{pmatrix} 1 & 0 & a \\ 0 & \mathbf{I} & A(a) \\ 0 & 0 & \mathbf{I} \end{pmatrix} \begin{pmatrix} \pi \\ \phi \\ \omega \end{pmatrix}$$

and hence

$$\dim \mathcal{G}^{(1)} = \dim\{a\} = n.$$

Now

$$\dim \mathcal{G}^{(1)} = \sigma$$

if and only if

$$n = n + \sum_{j=2}^{n} j(n-j)$$

or if and only if $n = 2$. Since we treated that case in Lecture 7 we now assume $n \geq 3$.

Now we prolong to $U \times G \times G^{(1)}$. The group

$$\begin{pmatrix} 1 & 0 & a \\ 0 & \mathbf{I} & A(a) \\ 0 & 0 & \mathbf{I} \end{pmatrix}$$

is Abelian and in fact a faithful representation of $\mathbf{R}^n$. The new Maurer–Cartan matrix is

$$\tilde{\omega}(S) = \begin{pmatrix} 0 & 0 & \alpha \\ 0 & 0 & \eta \\ 0 & 0 & 0 \end{pmatrix}$$

where explicitly

$$\alpha = da, \qquad \eta = dA(a).$$

Since $G^{(1)}$ was the group preserving the equations

$$d\omega = (\pi\mathbf{I} + \phi) \wedge \omega,$$

the lifts to $U \times G \times G^{(1)}$ satisfy the same equations. Therefore we will not add notation to indicate these lifts.

Next we need to study the derivatives of ϕ and π. Since

$$\begin{aligned} 0 = d^2\omega &= (d\pi\mathbf{I} + d\phi) \wedge \omega - (\pi\mathbf{I} + \phi) \wedge (\pi\mathbf{I} + \phi) \wedge \omega \\ &= (d\pi\mathbf{I} + d\phi - \phi \wedge \phi) \wedge \omega, \end{aligned}$$

the same argument given in Lecture 6, in the Riemannian geometry example, implies

$$(d\pi\mathbf{I} + d\phi - \phi \wedge \phi) = \mu \wedge \omega.$$

Taking $1/n$ trace we have the existence of a vector 1–form τ where

$$d\pi = \tau \wedge \omega.$$

Substituting this back into the equation for $d\pi$ yields

$$d\phi - \phi \wedge \phi = (\mu - \tau \mathbf{I}) \wedge \omega = \sigma \wedge \omega.$$

Now collecting this information together we have

$$d\begin{pmatrix} \pi \\ \phi \\ \omega \end{pmatrix} = \begin{pmatrix} 0 & 0 & \tau \\ 0 & 0 & \sigma \\ 0 & 0 & 0 \end{pmatrix} \wedge \begin{pmatrix} \pi \\ \phi \\ \omega \end{pmatrix} + \begin{pmatrix} 0 \\ \phi \wedge \phi \\ (\pi \mathbf{I} + \phi) \wedge \omega \end{pmatrix},$$

the exterior derivatives of the lifts of the forms from $U \times G$ to $U \times G \times G^{(1)}$. Note that because of integrability conditions, none of the zero blocks led to any torsion.

Next we need to find the principal components of $G^{(1)}$. Differentiation of the defining equation

$$A_i{}^j{}_k = a_i \delta^j_k - a_j \delta^i_k$$

gives

$$dA_i{}^j{}_k = da_i \delta^j_k - da_j \delta^i_k$$

or using our notation for the Maurer–Cartan forms

$$\eta_i{}^j{}_k - \alpha_i \delta^j_k + \alpha_j \delta^i_k = 0.$$

These are the obvious zero relations on the defining relations of $\mathcal{G}^{(1)}$. As such,

$$\sigma^i_{jk} - \tau_i \delta_{jk} + \tau_j \delta_{ik}$$

are the only principal components of order zero for the prolonged system.

Applying d to the structure equation for $d\omega$ gives

$$\begin{aligned} 0 = d(d\omega) &= d\pi \mathbf{I} \wedge \omega + d\phi \wedge \omega - (\pi \mathbf{I} + \phi) \wedge d\omega \\ &= (d\pi \mathbf{I} + d\phi - (\pi \mathbf{I} + \phi) \wedge (\pi \mathbf{I} + \phi)) \wedge \omega \\ &= (\tau \wedge \omega \mathbf{I} + d\phi - \phi \wedge \phi) \wedge \omega \end{aligned}$$

and by Cartan's lemma

$$d\phi^i_j - \phi^i_k \wedge \phi^k_j - (\tau_j \delta_{jk} + \tau_j \phi_{ik}) \wedge \omega^k = \sum \Delta^i_{jk} \wedge \omega^k$$

with

$$\Delta^i_{jk} = -\Delta^i_{kj}.$$

Using the same S_3 lemma argument that was used in the Riemannian geometry example in Lecture 6, we find the integrability conditions

$$\Delta^i_{jk} = \frac{1}{2} \sum w^i_{jkl}\, \omega^l.$$

Now we utilize the principal component decomposition and write

$$d\phi^i_j = \sum(\tau_i\delta_{jk} - \tau_j\delta_{ik}) \wedge \omega^k + \sum \phi^i_k \wedge \phi^k_j + \sum(\sigma^i_{jk} - \tau_i\delta_{jk} + \tau_j\delta_{ik}) \wedge \omega^k$$

and comparing with the integrability conditions we have

$$d\phi^i_j - \sum \phi^i_k \wedge \phi^k_j + (\tau_j\delta_{ik} - \tau_i\delta_{jk}) \wedge \omega^k = \frac{1}{2}\sum w^i_{jkl}\,\omega^l \wedge \omega^k.$$

The τ's are not yet unique since if

$$d\pi = \tau' \wedge \omega$$

we have

$$(\tau' - \tau) \wedge \omega = 0,$$

or by Cartan's lemma

$$\tau'_k = \tau_k + \sum b_{kl}\,\omega^l,$$

with $b_{kl} = b_{lk}$.

Next we want to solve the maximal number of equations of the form

$$\sum(\beta_j\delta_{il} - \beta_i\delta_{jl}) \wedge \omega^l = \frac{1}{2}\sum w^i_{jkl}\,\omega^k \wedge \omega^l,$$

where

$$\beta_j = \sum b_{jl}\,\omega^l.$$

If there were solutions then by Cartan's lemma

$$\sum \beta_j\,\delta_{il} - \beta_i\,\delta_{jl} - \frac{1}{2}w^i_{jkl}\omega^k = \sum k^i_{jlm}\,\omega^m,$$

and contraction with i and l give

$$n\beta_j - \beta_j - \frac{1}{2}\sum w^i_{jki}\,\omega^k = \sum k^i_{jlm}\,\omega^k.$$

Thus if we set

$$\beta_j = \frac{1}{2(n-1)}\sum w^i_{jkl}\,\omega^k$$

we achieve the normalization

$$\sum w^i_{jki} = 0$$

and make the τ_k unique, and $G^{(2)} = e$. This produces an e-structure on $U \times G \times G^{(1)}$.

Finally we need the form of the derivatives of the τ's. Since

$$\begin{aligned} d\pi &= \tau \wedge \omega \\ 0 = d^2\pi &= d\tau \wedge \omega - \tau \wedge d\omega \\ &= d\tau \wedge \omega - \tau \wedge (\pi \mathbf{I} + \phi) \wedge \omega \end{aligned}$$

and hence by Cartan's lemma

$$d\tau - \tau \wedge \pi - \tau \wedge \phi = \lambda \wedge \omega,$$

where ${}^t\lambda = \lambda$. Collecting all this information together we have the structure equations of conformal geometry

$$\begin{aligned} d\omega^i &= \pi \wedge \omega^i + \sum \phi^i_k \wedge \omega^k \\ d\phi^i_j &= \sum \phi^i_k \wedge \phi^k_j + (\tau_i \delta_{jk} - \tau_j \delta_{ik}) \wedge \omega^k - \frac{1}{2} \sum w^i_{jkl} \omega^l \wedge \omega^k \\ d\pi &= \sum \tau_i \wedge \omega^i \\ d\tau_k &= \tau_k \wedge \pi - \sum \tau_j \wedge \phi^j_k - \sum \lambda_{kl} \wedge \omega^l \end{aligned}$$

with

$$\sum w^i_{jil} = 0 \quad \text{and} \quad \lambda_{kl} = \lambda_{lk}.$$

Example 5. Third-order ordinary differential equations under contact transformations (continued). The details of the solution of this equivalence problem are too long to present, but by now the interest is in the overall structure of the problem, not in calculational technique.

The problem begins with a nine-dimensional solvable group G and structure equations

$$d \begin{pmatrix} \omega^1 \\ \omega^2 \\ \omega^3 \\ \omega^4 \end{pmatrix} = \begin{pmatrix} \alpha & 0 & 0 & 0 \\ \beta^1 & \beta^2 & 0 & 0 \\ \gamma^1 & \gamma^2 & \gamma^3 & 0 \\ \eta^1 & \eta^2 & 0 & \eta^3 \end{pmatrix} \wedge \begin{pmatrix} \omega^1 \\ \omega^2 \\ \omega^3 \\ \omega^4 \end{pmatrix} + \begin{pmatrix} A\,\omega^2 \wedge \omega^4 \\ B\,\omega^3 \wedge \omega^4 \\ 0 \\ 0 \end{pmatrix}.$$

The fiber action on torsion is

$$\left. \begin{aligned} dA - A\,\alpha + A\,\beta^2 + A\,\eta^3 \equiv 0 \\ dB - B\,\beta^2 + B\,\gamma^3 + B\,\eta^3 \equiv 0 \end{aligned} \right\} \quad \text{mod base}$$

and a parametric calculation gives $A \neq 0$ and $B \neq 0$. We normalize the torsion by setting

$$A = -1, \qquad B = -1.$$

The first-order normalizations yield a seven-dimensional group G_1 with new principal components

$$\alpha - \beta^2 - \eta^3, \qquad \beta^2 - \gamma^3 - \eta^3.$$

This leads to the structure equations

$$d\begin{pmatrix}\omega^1\\ \omega^2\\ \omega^3\\ \omega^4\end{pmatrix} = \begin{pmatrix}\alpha & 0 & 0 & 0\\ \beta^1 & \beta^2 & 0 & 0\\ \gamma^1 & \gamma^2 & 2\beta^2-\alpha & 0\\ \eta^1 & \eta^2 & 0 & \alpha-\beta^2\end{pmatrix} \wedge \begin{pmatrix}\omega^1\\ \omega^2\\ \omega^3\\ \omega^4\end{pmatrix} + \begin{pmatrix}-\omega^2\wedge\omega^4\\ -\omega^3\wedge\omega^4\\ A_1\,\omega^4\wedge\omega^3\\ B_1\,\omega^3\wedge\omega^4\end{pmatrix}.$$

Not all this torsion is really there, since the integrability condition $d^2\omega^1 = 0$ implies $B_1 = 0$. The fiber action on torsion is

$$dA_1 - A_1(\alpha - \beta^2) - 3\beta^1 + 3\gamma^2 \equiv 0$$

and we can translate A_1 to zero:

$$A_1 = 0.$$

The second-order normalizations yield a six-dimensional group G_2 with a new principal component, $\gamma^2 - \beta^1$.

This leads to the structure equations

$$d\begin{pmatrix}\omega^1\\ \omega^2\\ \omega^3\\ \omega^4\end{pmatrix} = \begin{pmatrix}\alpha & 0 & 0 & 0\\ \beta^1 & \beta^2 & 0 & 0\\ \gamma^1 & \beta^1 & 2\beta^2-\alpha & 0\\ \eta^1 & \eta^2 & 0 & \alpha-\beta^2\end{pmatrix} \wedge \begin{pmatrix}\omega^1\\ \omega^2\\ \omega^3\\ \omega^4\end{pmatrix} + \begin{pmatrix}-\omega^2\wedge\omega^4\\ -\omega^3\wedge\omega^4\\ A_2\,\omega^4\wedge\omega^2\\ 0\end{pmatrix}.$$

The fiber action on torsion is

$$dA_2 + 2A_2\,\alpha - 2A_2\,\beta^2 - 2\gamma^1 \equiv 0 \qquad \text{mod base},$$

and we can translate A_2 to zero:

$$A_2 = 0.$$

The third-order normalizations yield a five-dimensional group G_3 with a new principal component, γ^1.

This leads to the structure equations

$$d\begin{pmatrix}\omega^1\\ \omega^2\\ \omega^3\\ \omega^4\end{pmatrix} = \begin{pmatrix}\alpha & 0 & 0 & 0\\ \beta^1 & \beta^2 & 0 & 0\\ 0 & \beta^1 & 2\beta^2-\alpha & 0\\ \eta^1 & \eta^2 & 0 & \alpha-\beta^2\end{pmatrix} \wedge \begin{pmatrix}\omega^1\\ \omega^2\\ \omega^3\\ \omega^4\end{pmatrix} + \begin{pmatrix}-\omega^2\wedge\omega^4\\ -\omega^3\wedge\omega^4\\ A_3\,\omega^1\wedge\omega^4\\ 0\end{pmatrix}.$$

The fiber action on torsion is

$$dA_3 + 3A_3\,\alpha - 3A_3\,\beta^2 \equiv 0 \qquad \text{mod base},$$

and here we have a bifurcation in the flowchart depending on whether A_3 is zero. A parametric calculation shows that

$$A_3 = \frac{Ic^2}{a^3},$$

where

$$I = -f_y - \frac{1}{3} f_{y'} f_{y''} - \frac{2}{27}(f_{y''})^3 + \frac{1}{2}\frac{d}{dx}(f_{y'}) + \frac{1}{3} f_{y''} \frac{d}{dx}(f_{y'}) - \frac{1}{6}\frac{d^2}{dx^2}(f_{y''}).$$

Note this has the expected dependence on third derivatives.

Case I. $A_3 = 0$. The structure equations have the form

$$d\begin{pmatrix} \omega^1 \\ \omega^2 \\ \omega^3 \\ \omega^4 \end{pmatrix} = \begin{pmatrix} \alpha & 0 & 0 & 0 \\ \beta^1 & \beta^2 & 0 & 0 \\ 0 & \beta^1 & 2\beta^2 - \alpha & 0 \\ \eta^1 & \eta^2 & 0 & \alpha - \beta^2 \end{pmatrix} \wedge \begin{pmatrix} \omega^1 \\ \omega^2 \\ \omega^3 \\ \omega^4 \end{pmatrix} + \begin{pmatrix} -\omega^2 \wedge \omega^4 \\ -\omega^3 \wedge \omega^4 \\ 0 \\ 0 \end{pmatrix}$$

and there is no more reduction. The Cartan characters are $\sigma_1 = 3$, $\sigma_2 = 2$ and g is

$$3 + 2(2) = 7.$$

We compute $G_3^{(1)}$ and we see

$$\begin{aligned} \bar{\alpha} &= \alpha + 2p\omega^1 \\ \bar{\beta}^1 &= \beta^1 + p\omega^2 \\ \bar{\beta}^2 &= \beta^2 + p\omega^1 \\ \bar{\eta}^1 &= \eta^1 + q\omega^1 + r\omega^2 + p\omega^4 \\ \bar{\eta}^2 &= \eta^2 + r\omega^1 + s\omega^2 \end{aligned}$$

and hence

$$\dim G_3^{(1)} = 4.$$

The system is not in involution and we must prolong.

After some calculations there are three new torsion terms all of which can be translated to zero. This reduces the group to one dimension and a final normalization reduces the group to the identity. Thus the branch $A_3 = 0$ took three times around Loop A, prolonged and then twice more around Loop A. The result is an e-structure on a ten-dimensional space with five structure functions upon which equivalence depends.

Case II. $A_3 \neq 0$. The fiber action is

$$dA_3 - 3A_3\,\alpha - 3A_3\,\beta^2 \equiv 0 \qquad \text{mod base.}$$

This means A_3 is acted on by cubes and we can normalize $A_3 = 1$. The fourth-order normalization yields a four-dimensional group G_4 with a new principal component $\alpha - \beta^2$. This leads to the structure equation with three torsion terms A_4, B_4, C_4, and fiber action

$$\begin{aligned} &dA_4 - A_4\,\alpha - \eta^2 \equiv 0 \\ &dB_4 - \beta^1 \equiv 0 \qquad\qquad \text{mod base.} \\ &dC_4 + C_4\,\alpha + 2B_4\,\eta^2 + 2A_4\,\beta^1 \equiv 0 \end{aligned}$$

Now A_4 and B_4 can be translated to zero:

$$A_4 = 0, \qquad B_4 = 0,$$

and there is another bifurcation depending on the vanishing of C_4.

Case IIa. $C_4 = 0$. The reduction yields a two-dimensional group G_5 with principal components η^2, β^1. There are two more torsion terms with coefficients A_5, B_5, C_5, D_5, E_5, F_5, and there is a fiber action

$$dB_5 + B_5\,\alpha - \eta^1 \equiv 0 \qquad \text{mod base},$$

hence B_5 can be translated to zero which reduces us to a one-dimensional group G_6. The structure equations mod torsion now look like

$$d\begin{pmatrix}\omega^1\\ \omega^2\\ \omega^3\\ \omega^4\end{pmatrix} \equiv \begin{pmatrix}\alpha & 0 & 0 & 0\\ 0 & \alpha & 0 & 0\\ 0 & 0 & \alpha & 0\\ 0 & 0 & 0 & 0\end{pmatrix} \wedge \begin{pmatrix}\omega^1\\ \omega^2\\ \omega^3\\ \omega^4\end{pmatrix}$$

and we see α is now uniquely defined. This gives an e-structure on the five-dimensional space $U \times G_6$ with structure equations

$$\begin{aligned}
d\omega^1 &= \alpha \wedge \omega^1 - \omega^2 \wedge \omega^4\\
d\omega^2 &= \alpha \wedge \omega^2 - \omega^3 \wedge \omega^4 + \mathbf{a}\,\omega^2 \wedge \omega^1 + \mathbf{b}\,\omega^3 \wedge \omega^1 + \mathbf{c}\,\omega^4 \wedge \omega^1\\
d\omega^3 &= \alpha \wedge \omega^3 + \omega^1 \wedge \omega^4 + \mathbf{e}\,\omega^3 \wedge \omega^2 + \mathbf{f}\,\omega^1 \wedge \omega^2 + \mathbf{c}\,\omega^4 \wedge \omega^2\\
d\omega^4 &= \mathbf{g}\,\omega^3 \wedge \omega^2 + \mathbf{h}\,\omega^2 \wedge \omega^1 + \mathbf{i}\,\omega^3 \wedge \omega^1 + \mathbf{j}\,\omega^4 \wedge \omega^1\\
d\alpha &= \mathbf{k}\,\omega^1 \wedge \omega^2 + \mathbf{l}\,\omega^1 \wedge \omega^3 + (\mathbf{f} + cb - ce)\omega^1 \wedge \omega^4 \mathbf{i}\,\omega^2 \wedge \omega^3\\
&\qquad - \mathbf{b}\,\omega^3 \wedge \omega^4 - (\mathbf{a} + \mathbf{j})\omega^2 \wedge \omega^4.
\end{aligned}$$

The twelve functions $\mathbf{a}, \cdots, \mathbf{l}$ are the structure functions in terms of which equivalence is formulated.

Case IIb. $C_4 \neq 0$. In this case we can normalize C_4 to 1, which reduces us to a one-dimensional group G_6 with new principal component α. The resulting torsion has a coefficient which can be translated to zero, which results in an e-structure on the original four-dimensional space.

Now we take up the question of how to identify the group of automorphisms of an e-structure. This involves serious representation theory, and I will only outline the procedure assuming the Lie algebra is semisimple.

Let $\mathcal{G}$ be a semisimple Lie algebra and $\mathcal{H}$ a maximal Abelian subalgebra. General theory tells us that the cotangent space $T_e^*(G)$, when acted upon by the coadjoint representation, splits as a direct sum of $\mathcal{G}$–modules

$$\bigoplus_{\lambda} V_\lambda^*,$$

where

$$V_\lambda^* = \{\omega \in T_e^*(G) \mid \mathrm{ad}^\dagger(X)\,\omega = -\lambda(X)\,\omega \text{ for all } X \in \mathcal{H}\}.$$

The linear functions λ are called roots of $\mathcal{G}$ and the structure of $\mathcal{G}$ can be determined from the classification theorem of semisimple Lie algebras.

Example 10. *The invariants of a generic system of three equations in five variables.* These have a coframe ω satisfying

$$\begin{aligned} d\omega^1 &\equiv \omega^3 \wedge \omega^4 && \bmod(\omega^1, \omega^2) \\ d\omega^2 &\equiv \omega^3 \wedge \omega^5 && \bmod(\omega^1, \omega^2) \\ d\omega^3 &\equiv \omega^4 \wedge \omega^5 && \bmod(\omega^1, \omega^2, \omega^3) \end{aligned}$$

and gives rise to an equivalence problem needing two prolongations, resulting in an e-structure on a 14-dimensional space with structure equations [C. 1910]

$$\begin{aligned}
d\omega^1 &= \omega^1(2\varpi^1 + \varpi^4) + \omega^2\varpi^2 + \omega^3\omega^4 \\
d\omega^2 &= \omega^1\varpi^3 + \omega^2(\varpi^1 + 2\varpi^4) + \omega^3\omega^5 \\
d\omega^3 &= \omega^1\varpi^5 + \omega^2\varpi^6 + \omega^3(\varpi^1 + \varpi^4) + \omega^4\omega^5 \\
d\omega^4 &\equiv \omega^1\varpi^7 + \frac{4}{3}\omega^3\varpi^6 + \omega^4\varpi^1 + \omega^5\varpi^2 \\
d\omega^5 &= \omega^2\varpi^7 - \frac{4}{3}\omega^3\varpi^5 + \omega^4\varpi^3 + \omega^5\varpi^4 \\
d\varpi^1 &= \varpi^3\varpi^2 + \frac{1}{3}\omega^3\varpi^7 - \frac{2}{3}\omega^4\varpi^5 + \frac{1}{3}\omega^5\varpi^6 + \omega^1\chi^1 + 2B_2\omega^1\omega^3 \\
&\quad + B_3\omega^2\omega^3 + 2A_2\omega^1\omega^4 + 2A_3\omega^1\omega^5 + A_3\omega^2\omega^4 + A_4\omega^2\omega^5 \\
d\varpi^2 &= \varpi^2(\varpi^1 - \varpi^4) - \omega^4\varpi^6 + \omega^1\chi^2 + B_4\omega^2\omega^3 + A_4\omega^2\omega^4 + A_5\omega^2\omega^5 \\
d\varpi^3 &= \varpi^3(\varpi^4 - \varpi^1) - \omega^5\varpi^5 + \omega^2\chi^1 - B_1\omega^1\omega^3 - A_1\omega^1\omega^4 - A_2\omega^1\omega^5 \\
d\varpi^4 &= \varpi^2\varpi^3 + \frac{1}{3}\omega^3\varpi^7 + \frac{1}{3}\omega^4\varpi^5 - \frac{2}{3}\omega^5\varpi^6 + \omega^2\chi^2 - B_2\omega^1\omega^3 \\
&\quad - 2B_3\omega^2\omega^3 - A_2\omega^1\omega^4 - A_3\omega^1\omega^5 - 2A_3\omega^2\omega^4 - 2A_4\omega^2\omega^5 \\
d\varpi^5 &= \varpi^1\varpi^5 + \varpi^3\varpi^6 - \omega^5\varpi^7 + \omega^3\chi^1 + \frac{9}{32}D_1\omega^1\omega^2 + \frac{9}{8}C_1\omega^1\omega^3 \\
&\quad + \frac{9}{8}C_2\omega^2\omega^3 + A_2\omega^3\omega^4 + A_3\omega^3\omega^5 + \frac{3}{4}B_1\omega^1\omega^4 \\
&\quad + \frac{3}{4}B_2(\omega^1\omega^5 + \omega^2\omega^4) + \frac{3}{4}B_3\omega^2\omega^5 \\
d\varpi^6 &= \varpi^2\varpi^5 + \varpi^4\varpi^6 + \omega^4\varpi^7 + \omega^3\chi^2 + \frac{9}{32}D_2\omega^1\omega^2 + \frac{9}{8}C_2\omega^1\omega^3 \\
&\quad + \frac{9}{8}C_3\omega^2\omega^3 - A_3\omega^3\omega^4 - A_4\omega^3\omega^5 + \frac{3}{4}B_2\omega^1\omega^4 \\
&\quad + \frac{3}{4}B_3(\omega^1\omega^5 + \omega^2\omega^4) + \frac{3}{4}B_4\omega^2\omega^5 \\
d\varpi^7 &= \frac{4}{3}\varpi^5\varpi^6 + (\varpi^1 + \varpi^4)\varpi^7 + \omega^4\chi^1 + \omega^5\chi^2 + \frac{9}{64}E\omega^1\omega^2 - \frac{3}{8}D_1\omega^1\omega^3 \\
&\quad - \frac{3}{8}D_2\omega^2\omega^3 + 2A_3\omega^4\omega^5 - B_2\omega^3\omega^4 + B_3\omega^3\omega^5
\end{aligned}$$

$$d\chi^1 \equiv \varpi^5\varpi^7 + (2\varpi^1 + \varpi^4)\chi^1 + \varpi^3\chi^2 \quad \mathrm{mod}(\omega^1, \cdots, \omega^5)$$
$$d\chi^2 \equiv \varpi^6\varpi^7 + \varpi^2\chi^1 + (\varpi^1 + 2\varpi^4)\chi^2 \quad \mathrm{mod}(\omega^1, \cdots, \omega^5).$$

The largest group will occur when all the 15 torsion coefficients are constant and in fact zero, and by our work we know that the resulting 14-dimensional space is locally a group.

If we let $\{X_i\}$ be dual to $\{\omega^i\}$, $\{Y_\rho\}$ dual to $\{\varpi^\rho\}$, and $\{Z_a\}$ dual to $\{\chi^a\}$, then $\{Y_1, Y_4\}$ acts diagonally and

$$\lambda_1(\pi^1) = \lambda_1(\pi^4) = \lambda_4(\pi^1) = \lambda_4(\pi^4) = 0,$$

hence $\{Y_1, Y_4\}$ is Abelian, and in fact maximal Abelian.

Now we compute $\lambda_1(\alpha)$ and $\lambda_4(\alpha)$ for the remaining α in the e-structure and plot

$$\alpha \to (\lambda_1(\alpha), \lambda_4(\alpha))$$

in the plane $\mathbf{R}^2$.

The roots can be read off from the torsion-free part of the structure equations by sight. Consider

$$d\omega^1 = \omega^1 \wedge (2\varpi^1 + \varpi^4) + \omega^1 \wedge \varpi^2$$

then

$$\mathrm{ad}(Y_1)\,\omega^1 = Y_1 \lrcorner d\omega^1 = -2\omega^1$$

and

$$\mathrm{ad}(Y_4)\,\omega^1 = Y_4 \lrcorner d\omega^1 = -\omega^1.$$

Thus

$$\omega^1 \to (2, 1).$$

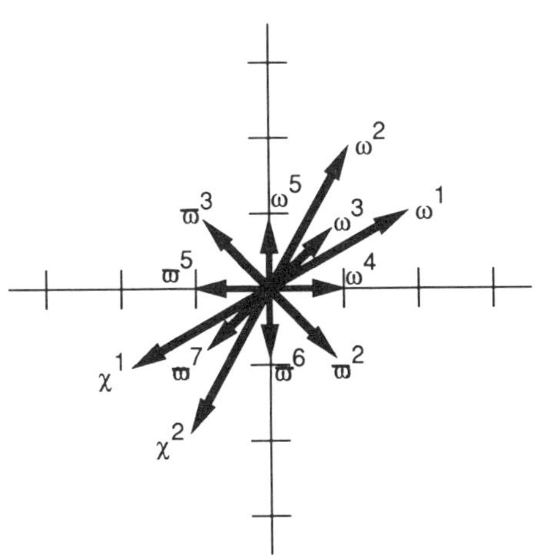

	ϖ^1	ϖ^4
ω^1	2	1
ω^2	1	2
ω^3	1	1
ω^4	1	0
ω^5	0	1
ϖ^1	0	0
ϖ^2	1	-1
ϖ^3	-1	1
ϖ^4	0	0
ϖ^5	-1	0
ϖ^6	0	-1
ϖ^7	-1	-1
χ^1	-2	-1
χ^2	-1	-2

FIG. 5

After seeing the computation you see that the dual roots are just the coefficients of the terms in $d\alpha$ of the form $\alpha \wedge \varpi^1$ and $\alpha \wedge \varpi^4$. After reading off and plotting the roots we find the diagram in Fig. 5.

This is an affine dual root diagram of the split form of the exceptional simple Lie group of dimension 14, called G_2. This is the noncompact real form of G_2.

Finally, after ten lectures on the method of equivalence, the whole subject can be looked at as the implementation of the flowchart in Fig. 6.

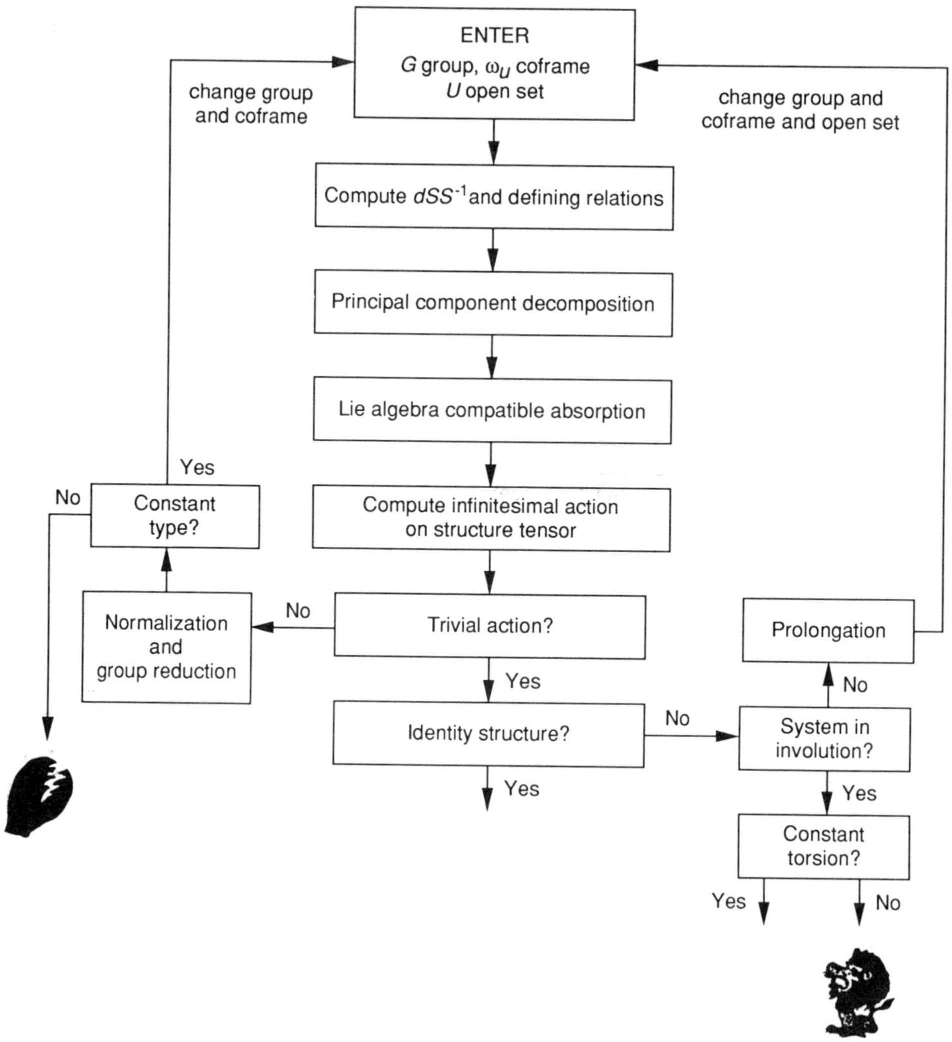

FIG. 6. *Method of equivalence for first-order determined problems.*

Appendix

Translation of Chapter 5, §§1–12 of É. Cartan's paper "Les sous-groupes des groupes continus de transformations," *Annales de l'École Normale* (3), 25 (1908), pp. 57-76.

1. Let us consider two systems of n expressions of independent linear total differentials in n variables: one in the n variables $x_1, x_2, \cdots, x_n$,

$$(1)\qquad \begin{cases} \omega_1 = a_{11}dx_1 + a_{12}dx_2 + \cdots + a_{1n}dx_n\,, \\ \omega_2 = a_{21}dx_1 + a_{22}dx_2 + \cdots + a_{2n}dx_n\,, \\ \cdots\cdots\cdots\cdots\cdots\cdots\cdots\cdots\cdots\,, \\ \omega_n = a_{n1}dx_1 + a_{n2}dx_2 + \cdots + a_{nn}dx_n; \end{cases}$$

the other in the n variables $X_1, X_2, \cdots, X_n$,

$$(2)\qquad \begin{cases} \Omega_1 = A_{11}dX_1 + A_{12}dX_2 + \cdots + A_{1n}dX_n\,, \\ \Omega_2 = A_{21}dX_1 + A_{22}dX_2 + \cdots + A_{2n}dX_n\,, \\ \cdots\cdots\cdots\cdots\cdots\cdots\cdots\cdots\cdots\,, \\ \Omega_n = A_{n1}dX_1 + A_{n2}dX_2 + \cdots + A_{nn}dX_n; \end{cases}$$

the a_{ik} denote given functions of the x's and the A_{ik} denote given functions of the X's. The problem at hand is that of recognizing whether we can find for the X's independent functions of the x's such that $\Omega_1, \Omega_2, \cdots, \Omega_n$ reduce respectively to $\omega_1, \omega_2, \cdots, \omega_n$.

The solution which will be exhibited is based on the consideration of the bilinear covariants.[1] If we calculate the n covariants of $\omega_1, \omega_2, \cdots, \omega_n$ we can put them in the form

$$(3)\qquad \omega'_s = \sum_{(ik)}^{1,2,\cdots,n} c_{iks}\omega_i\omega_k \qquad (s = 1, 2, \cdots, n),$$

and similarly the covariants Ω' are in the form

$$(4)\qquad \Omega'_s = \sum_{(ik)}^{1,2,\cdots,n} C_{iks}\Omega_i\Omega_k \qquad (s = 1, 2, \cdots, n).$$

[1]See É. Cartan, *Annales de l'École Normale* (3), 21 (1904), pp. 154-156.

If a change of variables transforms the ω_s's into the Ω_s's, it will also transform the ω_s''s into the Ω_s''s, and consequently we have

$$(5) \qquad C_{iks} = c_{iks}.$$

Let us suppose that among the $n^2(n-1)/2$ coefficients c_{iks}, r are independent, the others being functions of these.

Let us denote them by $y_1, y_2, \cdots, y_r$. We can write the differentials dy_i in the following form:

$$(6) \qquad \begin{cases} dy_1 = h_{11}\,\omega_1 + h_{12}\,\omega_2 + \cdots + h_{1n}\,\omega_n, \\ dy_2 = h_{21}\,\omega_1 + h_{22}\,\omega_2 + \cdots + h_{2n}\,\omega_n, \\ \cdots\cdots\cdots\cdots\cdots\cdots\cdots\cdots\cdots\cdots\cdots, \\ dy_r = h_{r1}\,\omega_1 + h_{r2}\,\omega_2 + \cdots + h_{rn}\,\omega_n. \end{cases}$$

Let us suppose that among the coefficients h there are $r_1 - r$ that are independent among themselves and also independent of the y's, denoted by $y_{r+1}, y_{r+2}, \cdots, y_{r_1}$. We calculate their differentials:

$$(7) \qquad \begin{cases} dy_{r+1} = h_{r+1,1}\,\omega_1 + h_{r+1,2}\,\omega_2 + \cdots + h_{r+1,n}\,\omega_n, \\ \cdots\cdots\cdots\cdots\cdots\cdots\cdots\cdots\cdots\cdots\cdots\cdots\cdots, \\ dy_{r_1} = h_{r_1 1}\omega_1 + h_{r_1 2}\omega_2 + \cdots + h_{r_1 n}\omega_n. \end{cases}$$

If among the new coefficients h there are $r_2 - r_1$ that are independent among themselves and also independent of the y's, denoted by $y_{r_1+1}, y_{r_1+2}, \cdots, y_{r_2}$, we calculate their differentials and repeat this process until we arrive at no new independent functions.

We will have, finally, obtained thus a certain number $p \leq n$ of independent functions of the x's, denoted by $y_1, y_2, \cdots, y_p$, having the following two properties:

1^0 The coefficients c_{iks} of formula (3) are functions of the y's;

2^0 The coefficients of the differentials dy expressed linearly in $\omega_1, \omega_2, \cdots, \omega_n$ are also functions of the y's.

2. This being done, if the transformation from system (1) to system (2) is possible, *it is necessary that the same operations applied to system* (2) *give rise to the same number of functions Y, and moreover that the coefficients C_{iks} in formulas* (4) *and the coefficients H_{ik} of the Ω's in the expansion of the dY_i be the same functions of the Y's as the corresponding quantities c_{iks} and h_{ik} were functions of the y's.*

This necessary condition is also sufficient and we will see that the most general transformation, which permits us to pass from system (1) to system (2), depends on $n - p$ arbitrary constants.

In fact let us suppose, to fix ideas, that the p differentials dy are linearly independent in the $\omega_1, \omega_2, \cdots, \omega_n$, and let us consider the system

$$(8) \qquad \begin{cases} Y_1 - y_1 = 0, \\ Y_2 - y_2 = 0, \\ \cdots\cdots\cdots\cdots, \\ Y_p - y_p = 0, \\ \Omega_{p+1} - \omega_{p+1} = 0, \\ \cdots\cdots\cdots\cdots, \\ \Omega_n - \omega_n = 0. \end{cases}$$

It results from the hypothesis which will be made that this system implies

$$\Omega_1 = \omega_1,$$
$$\cdots\cdots\cdots,$$
$$\Omega_p = \omega_p$$

and that it suffices to integrate it to have the most general solution to the problem.

Therefore, bearing in mind the equality of the Y's and the y's, the bilinear covariants of the first terms $\Omega_{p+1} - \omega_{p+1}, \cdots, \Omega_n - \omega_n$ of the equations of this system are zero if we recall (8).

This system is thus completely integrable and the general solution depends on $n - p$ arbitrary constants.

If we have a particular transformation from system (1) to system (2), the most general transformation is obtained by operating on the x's with the most general transformation leaving invariant the expressions $\omega_1, \omega_2, \cdots, \omega_n$. This transformation generates a group with $n-p$ parameters of which the *structure* is defined by formulas (3), with invariants $y_1, y_2, \cdots, y_p$.[2] *This structure,* by a theorem proven elsewhere,[3] *is also defined by the constants* $c_{p+i,p+j,p+s} (i, j, s = 1, 2, \cdots, n - p)$ *where the x's are given arbitrary numerical values.*

3. It may happen that we are investigating the equivalence of the two systems (1) and (2) under the added restriction that a certain number h of given independent functions $z_1, z_2, \cdots, z_h$ of the x's transform into the same number h of given functions $Z_1, Z_2, \cdots, Z_h$ of the X's. Nothing changes in the solution, except that the given functions of the x's must be treated as the functions y in question in the general solution; one takes for $y_1, y_2, \cdots, y_r$ the h functions z and $r - h$ coefficients c_{iks} which are independent among themselves and independent of the z's, assuming that all the others may be expressed in terms of the z's and the $r - h$ coefficients c.

4. Let us now consider the following general problem:

Given on the one hand a system of n expressions of linearly independent total differentials $\omega_1, \omega_2, \cdots, \omega_n$ in $x_1, x_2, \cdots, x_n$, and m independent functions $y_1, y_2, \cdots, y_m$ of the n variables x; on the other hand, a system of n expressions

[2]See É. Cartan, *Annales de l'École Normale* (3), 21 (1904), p. 186.

[3]See É. Cartan, loc cit, p. 201.

of linearly independent total differentials $\Omega_1, \Omega_2, \cdots, \Omega_n$ *in* $X_1, X_2, \cdots, X_n$, *and m independent functions* $Y_1, Y_2, \cdots, Y_m$ *of these n variables X; determine if there exists a change of variables transforming respectively the functions* $y_1, \cdots, y_m$ *into the functions* $Y_1, \cdots, Y_m$, *also such that by this change of variables* $\Omega_1, \Omega_2, \cdots, \Omega_n$ *reduce to* $\omega_1, \omega_2, \cdots, \omega_n$ *by a linear substitution belonging to a given linear group with r parameters* Γ, *the coefficients of the finite equations of this group possibly depending on* $y_1, y_2, \cdots, y_m$.

In short, we must have relations of the form

$$(9) \qquad \begin{cases} \Omega_1 = \alpha_{11}(y,u)\,\omega_1 + \alpha_{12}(y,u)\,\omega_2 + \cdots + \alpha_{1n}(y,u)\,\omega_n\,, \\ \Omega_2 = \alpha_{21}(y,u)\,\omega_1 + \alpha_{22}(y,u)\,\omega_2 + \cdots + \alpha_{2n}(y,u)\,\omega_n\,, \\ \cdots\cdots\cdots\cdots\cdots\cdots\cdots\cdots\cdots\cdots\cdots\cdots, \\ \Omega_n = \alpha_{n1}(y,u)\,\omega_1 + \alpha_{n2}(y,u)\,\omega_2 + \cdots + \alpha_{nn}(y,u)\,\omega_n, \end{cases}$$

the α's being given functions of the $y_1, y_2, \cdots, y_n$ and the r quantities $u_1, u_2, \cdots, u_r$ which, for each change of variables solving the problem, will be certain functions of $x_1, x_2, \cdots, x_n$; furthermore, the functions α are such that, if one regards the u's as parameters and the y's as constants, the formulas (9) define the linear group Γ (on n variables $\omega_1, \omega_2, \cdots, \omega_n$).

We can formulate the problem in another way. Let us regard the u's as r new auxiliary variables, and let us suppose

$$(10) \quad \overline{\omega_s} = \alpha_{s1}(y,u)\,\omega_1 + \alpha_{s2}(y,u)\,\omega_2 + \cdots + \alpha_{sn}(y,u)\,\omega_n \qquad (s = 1, 2, \cdots, n).$$

Let us similarly suppose

$$(11) \quad \overline{\Omega_s} = \alpha_{s1}(Y,U)\,\Omega_1 + \alpha_{s2}(Y,U)\,\Omega_2 + \cdots + \alpha_{sn}(Y,U)\,\Omega_n \qquad (s = 1, 2, \cdots, n).$$

The original problem may be replaced by the following one:

To determine for $X_1, X_2, \cdots, X_n; U_1, \cdots, U_r$ *a system of* $n + r$ *independent functions of* $x_1, x_2, \cdots, x_n; u_1, \cdots, u_r$ *such that the functions Y become equal to the corresponding functions y and such that the expressions* $\overline{\Omega}$ *become equal to the corresponding functions* $\overline{\omega}$.

That given, *we are led back to the study of the invariants of the system formed by the m functions y and the n expressions* $\overline{\omega}$.

5. Let us form the differentials $dy_1, dy_2, \cdots, dy_m$ and express them in terms of $\overline{\omega_1}, \overline{\omega_2}, \cdots, \overline{\omega_n}$, thus

$$(12) \qquad \begin{cases} dy_1 = h_{11}\,\overline{\omega_1} + h_{12}\,\overline{\omega_2} + \cdots + h_{1n}\,\overline{\omega_n}\,, \\ \cdots\cdots\cdots\cdots\cdots\cdots\cdots\cdots, \\ dy_m = h_{m1}\overline{\omega_1} + h_{m2}\overline{\omega_2} + \cdots + h_{mn}\overline{\omega_n}. \end{cases}$$

The coefficients h_{ij}, functions of the x's and the u's, are covariants of the system under consideration.

Imagine that we operate on the $\overline{\omega}$'s with a substitution from the group Γ; the h_{ij} will undergo among themselves a linear substitution generating a group Γ' (dual to the group Γ). That is to say, if we designate by h_{ij}^0 the function of $x_1, x_2, \cdots, x_n$ to which h_{ij} reduces when we give to the u's the values corresponding to the identity transformation of Γ, the h_{ij} reduce to the h_{ij}^0 by the transformations of the group Γ'. The two systems of quantities h_{ij}^0 and h_{ij} are *homologous* under the group Γ'. That done, we can find a particular transformation of Γ' such that the system h_{ij}^0 becomes a *particular* homologous system; for example, if among the h's, considered as functions of the u's, there exist $r-r'$ independent ones, one can give to these $r-r'$ functions *fixed numerical values,* the others becoming functions of the x's which will be expressible in terms of $y_1, y_2, \cdots, y_m$ and of $m' - m$ among themselves, which we designate by $y_{m+1}, \cdots, y_{m'}$. These $m' - m$ new functions are the *invariants* of the system.

In sum, we have just found as invariant elements the integers r' and m' and the functions by which the coefficients h_{ij} are expressed in terms of $y_1, \cdots, y_m$, once $r - r'$ of them have been reduced to fixed numerical values.

That done, *we can always restrict ourselves to choosing the expressions ω on one hand, the expressions Ω on the other, in a manner such that the $r - r'$ coefficients h_{ij} can be considered to take the given numerical values given in advance* (if necessary, this is realized by replacing the expressions ω by others which result from a particular linear substitution of the group Γ, which changes none of the conditions of the problem).

Given that, *the most general linear transformation of Γ under which the ω's become the Ω's corresponds to the most general transformation of the group Γ' which leaves the system h_{ij}^0 invariant.* This transformation obviously generates a subgroup of Γ' with r' parameters. The h_{ij} being functions of $y_1, \cdots, y_m; u_1, \cdots, u_r$ (since they are obtained from the h_{ij}^0 by the general transformation of the group Γ'), this subgroup is defined by $r - r'$ relations between $u_1, u_2, \cdots, u_r, y_1, \cdots, y_{m'}$.

We are thus led back to the original problem, except that we have now $m' \geq m$ functions y and, instead of the group Γ, one of its subgroups with r' parameters, the coefficients of the finite equations of this subgroup depending on $y_1, y_2, \cdots, y_m$.

We will approach this new problem as we did the previous one until neither of the integers m and r undergo modification. This amounts in effect to the assumption that the coefficients h_{ij} in formulas (12) depend only on $y_1, y_2, \cdots, y_m$, and not at all on $u_1, u_2, \cdots, u_r$.

6. This first reduction of the problem having been performed, let us form the bilinear covariants of the expressions $\overline{\omega_s}$. The formulas (10) give

$$
\begin{aligned}
\overline{\omega_s'} = {} & \alpha_{s1}(y,u)\,\omega_1' + \alpha_{s2}(y,u)\,\omega_2' + \cdots + \alpha_{sn}(y,u)\,\omega_n' \\
& + \sum_i^{1,\cdots,m} dy_i \left(\frac{\partial \alpha_{s1}}{\partial y_i}\omega_1 + \frac{\partial \alpha_{s2}}{\partial y_i}\omega_2 + \cdots + \frac{\partial \alpha_{sn}}{\partial y_i}\omega_n \right) \\
& + \sum_i^{1,\cdots,r} du_k \left(\frac{\partial \alpha_{s1}}{\partial u_k}\omega_1 + \frac{\partial \alpha_{s2}}{\partial u_k}\omega_2 + \cdots + \frac{\partial \alpha_{sn}}{\partial u_k}\omega_n \right).
\end{aligned}
\tag{13}
$$

Therefore the covariants $\overline{\omega_i'}$ are expressible linearly in terms of the dx's, that is to say again in terms of the ω's and finally in terms of the $\overline{\varpi}$'s; it is the same with the dy_i's. Furthermore, let

$$\sum_{\rho}^{1,\cdots,r}\sum_{i,s}^{1,\cdots,n} e_\rho b_{i\rho s}\omega_i \frac{\partial f}{\partial \omega_s}$$

be the most general infinitesimal transformation of Γ, the $b_{i\rho s}$ being functions of $y_1, y_2, \cdots, y_m$. We have, as we know,

$$\frac{\partial \overline{\omega_s}}{\partial u_k} = \frac{\partial \alpha_{s1}}{\partial u_k}\omega_1 + \cdots + \frac{\partial \alpha_{sn}}{\partial u_k}\omega_n = \sum \lambda_{k\rho}(u)\, b_{i\rho s}\,\overline{\omega_i}.$$

Consequently the covariant $\overline{\omega_s'}$ can be put in the form

$$\overline{\omega_s'} = \sum_{(ik)} a_{iks}\,\overline{\omega_i}\,\overline{\omega_k} - \sum_{i,\rho} b_{i\rho s}\,\overline{\omega_i}\sum \lambda_{k\rho} du_k,$$

or also in the form

$$\text{(14)} \qquad \overline{\omega_s'} = \sum_{(ik)} a_{iks}\,\overline{\omega_i}\,\overline{\omega_k} + \sum_{i,\rho} b_{i\rho s}\,\overline{\omega_i}\,\varpi_\rho \qquad (s = 1, 2, \cdots, n),$$

the a_{ik} designating certain functions of the x's and the u's, the ϖ_ρ designating r differential expressions in the variables x and u, linearly independent in $du_1, du_2, \cdots, du_r$.

Let us suppose that a change of variables, which gives for the X's functions of the x's and for the U's functions of the x's and the u's, chosen as above, transforms the ω_s into Ω_s. We will have formulas

$$\overline{\Omega_s'} = \sum_{(ik)} A_{iks}\,\overline{\Omega_i}\,\overline{\Omega_k} + \sum b_{i\rho s}\,\overline{\Omega_i}\,\Pi_\rho$$

from which we get

$$\sum (A_{iks} - a_{iks})\,\overline{\omega_i}\,\overline{\omega_k} + \sum b_{i\rho s}\,\overline{\omega_i}\,(\Pi_\rho - \varpi_\rho) = 0 \qquad (s = 1, 2, \cdots, r).$$

These n equalities are possible only if we have formulas of the form

$$\text{(15)} \qquad \Pi_\rho = \varpi_\rho + \nu_{\rho 1}\overline{\omega_1} + \nu_{\rho 2}\overline{\omega_2} + \cdots + \nu_{\rho n}\overline{\omega_n} \qquad (\rho = 1, 2, \cdots, r),$$

$$\text{(16)} \qquad A_{iks} = a_{iks} + \sum_{\rho}^{1,\cdots,r} (b_{k\rho s}\nu_{\rho i} - b_{i\rho s}\nu_{\rho k}),$$

where the $\nu_{\rho i}$ designate appropriately chosen functions of the x's and the u's.

Let us then suppose that

$$(17)\qquad \begin{cases} \overline{\varpi_\rho} = \varpi_\rho + \nu_{\rho 1}\,\overline{\omega_1} + \cdots + \nu_{\rho n}\,\overline{\omega_n}, \\ \overline{a_{iks}} = a_{iks} + \sum\limits_{\rho}^{1,\cdots,r} (b_{k\rho s}\,\nu_{\rho i} - b_{i\rho s}\,\nu_{\rho k}), \end{cases}$$

where the $\nu_{\rho i}$ designate now the new auxiliary variables.

Given any change of variables whatsoever from the x's to the X's, we can determine nr quantities $\nu_{\rho i}$ such that we have

$$\overline{\Omega_s} = \overline{\omega_s}, \qquad \overline{\Pi_\rho} = \overline{\varpi_\rho}, \qquad \overline{A_{iks}} = \overline{a_{iks}}.$$

We are thus led back to studying the invariants of the system of the expressions $\overline{\omega_s}$ and $\overline{\varpi_\rho}$, the functions y_i, and the coefficients $\overline{a_{iks}}$.

7. Concerning these last coefficients, we will perform a reduction of the problem analogous to that which was performed earlier (no. 5). Let us operate on the $\overline{\omega}$'s with an infinitesimal transformation of the group Γ; that is

$$\delta\overline{\omega_s} = \sum_\rho \sum_i e_\rho\, b_{i\rho s}\,\overline{\omega_i}.$$

The formulas

$$\overline{\omega_s'} = \sum \overline{a_{iks}}\,\overline{\omega_i}\,\overline{\omega_k} + \sum b_{i\rho s}\,\overline{\omega_i}\,\overline{\varpi_\rho}$$

give us

$$\delta\overline{\omega_s'} = \sum \delta\overline{a_{iks}}\,\overline{\omega_i}\,\overline{\omega_k} + \sum \overline{a_{iks}}\,(\delta\,\overline{\omega_i}\,\overline{\omega_k} + \overline{\omega_i}\delta\overline{\omega_k}) + \sum b_{i\rho s}\,(\delta\overline{\omega_i}\,\overline{\varpi_\rho} + \overline{\omega_i}\,\delta\overline{\varpi_\rho}).$$

Therefore we demonstrate easily that the change in $\overline{\omega_s'}$ is equal to the bilinear covariant of $\delta\omega_s$:

$$\delta\,\overline{\omega_s'} = \sum de_\rho b_{i\rho s}\,\overline{\omega_i} + \sum e_\rho db_{i\rho s}\,\overline{\omega_i} + \sum e_\rho b_{i\rho s}\,\overline{\omega_i'}.$$

Equating the two values found for $\delta\overline{\omega_s'}$, and replacing the $\overline{\omega_i}, \delta\overline{\omega_i}, \cdots$ with their values, we obtain

$$\begin{aligned} (18)\qquad &\sum_{\lambda,\rho} b_{\lambda\rho s}\,\overline{\omega_\lambda}\,(\delta\,\overline{\varpi_\rho} + de_\rho) + \sum_{\rho,\lambda,\sigma} e_\rho \sum_i (b_{\lambda\rho i}\, b_{i\sigma s} - b_{\lambda\sigma i}\, b_{i\rho s})\,\overline{\omega_\lambda}\,\overline{\varpi_\sigma} \\ &+ \sum_{\lambda,\mu} \Bigg\{ \delta\,\overline{a_{\lambda\mu s}} + \sum_\rho e_\rho \Bigg[\sum_i (\overline{a_{i\mu s}}\, b_{\lambda\rho i} + \overline{a_{\lambda i s}}\, b_{\mu\rho i} - \overline{a_{\lambda\mu i}}\, b_{i\rho s}) \\ &\qquad + \sum_k \left(\frac{\partial b_{\lambda\rho s}}{\partial y_k}\, h_{k\mu} - \frac{\partial b_{\mu\rho s}}{\partial y_k}\, h_{k\lambda} \right) \Bigg] \Bigg\}\, \overline{\omega_\lambda}\,\overline{\omega_\mu} = 0. \end{aligned}$$

Therefore, if we introduce the structure constants $c_{\rho\sigma\tau}$ of the group Γ, we have

$$\sum_i (b_{\lambda\rho i}\, b_{i\sigma s} - b_{\lambda\sigma i}\, b_{i\rho s}) = \sum_\tau c_{\rho\sigma\tau}\, b_{\lambda\tau s}$$

and the first line of equation (18) reduces to

$$\sum_{\lambda,\rho} b_{\lambda\rho s}\,\overline{\varpi_\lambda}\left(\delta\,\overline{\varpi_\rho} + de_\rho - \sum_{\sigma,\tau} e_\tau c_{\sigma\tau\rho}\,\overline{\varpi_\sigma}\right).$$

We deduce from this the following formulas, where the $e_{\rho i}$ are nr arbitrary quantities:

$$(19)\quad \begin{cases} \delta\,\overline{\varpi_\rho} = -de_\rho - \sum_{\sigma,\tau} e_\sigma c_{\sigma\tau\rho}\overline{\varpi_\tau} + \sum_i e_{\rho i}\,\overline{\omega_i}, \\ \delta\,\overline{a_{\lambda\mu s}} = \sum_\rho e_\rho \left[\sum_i (\overline{a_{\lambda\mu i}}\, b_{i\rho s} - \overline{a_{\lambda i s}}\, b_{\mu\rho i} + \overline{a_{\mu i s}}\, b_{\lambda\rho i}) \right. \\ \qquad \left. + \sum_k \left(\dfrac{\partial b_{\mu\rho s}}{\partial y_k} h_{k\lambda} - \dfrac{\partial b_{\lambda\rho s}}{\partial y_k} h_{k\mu}\right)\right] + \sum_\rho (b_{\mu\rho s}\, e_{\rho\lambda} - b_{\lambda\rho s}\, e_{\rho\mu}). \end{cases}$$

The last formulas derived show that the coefficients $\overline{a}_{iks}$ are subject to a group Γ_1 with $r(n+1)$ parameters (of which the coefficients may depend on the y's). The quantities u_ρ and $\nu_{\rho i}$ are the parameters of the finite equations of this group. In other words, if we designate by $\overline{a^0_{iks}}$ that which a_{iks} becomes for particular values of the u's and the ν's, the general expressions $\overline{a_{iks}}$ reduce to $\overline{a^0_{iks}}$ by the most general transformation of the group Γ_1.

We deduce from this that if among the a_{iks} there are l that are independent when considered as functions of the u's and the ν's, we can suppose that the ω's are chosen in such a manner that the l coefficients have fixed numerical values given in advance; the others will then be *invariant* functions of the x's as expressed in terms of $y_1, y_2, \cdots, y_m$ and of an additional $m' - m$, denoted by $y_{m+1}, \cdots, y_{m'}$. Furthermore, *the group* Γ_1 *will reduce to the subgroup which conserves the* l *coefficients* $\overline{a_{iks}}$ *considered as given numerical values.*

Two cases are possible. Γ being hemihedrally isomorphic to Γ_1, it is possible that the reduction of Γ_1 will not result in Γ, and it is also possible that the reduction of Γ_1 will result in Γ. In the second case, from the relations between the u's, the ν's, and the y's which define the subgroup of Γ, we can deduce the relations between the u's and the y's only; these are the relations which define the subgroup of Γ to which one is led if m' is greater than m; the consideration of the invariants $y_{m+1}, \cdots, y_{m'}$ can also result in the group Γ.

We see thus that step by step we are led to the case where *all the coefficients* $\overline{a_{iks}}$ *are functions of* $y_1, y_2, \cdots, y_m$ *only, but depend neither on the* u*'s nor on the* ν*'s.* In general, the nr quantities $\nu_{\rho i}$ are not arbitrary; for example $nr - r_1$ among them can be expressed in terms of the y's, the u's, and the remaining r_1. The new group Γ_1 will be a group with $r + r_1$ parameters. The r expressions $\overline{\varpi_\rho}$ will depend only on the x's, the u's, and r_1 remaining parameters ν. The $nr - r_1$ linear relations between the e_ρ and the $e_{\rho i}$ may be obtained by expressing that the $\delta\overline{a_{\lambda\mu s}}$ are zero, that is to say

$$\sum_\rho (b_{\mu\rho s}e_{\rho\lambda} - b_{\lambda\rho s}e_{\rho\mu}) + \sum_\rho e_\rho \sum_i (\overline{a_{\lambda\mu i}}\, b_{i\rho s} - \overline{a_{\lambda i s}}\, b_{\mu\rho i} + \overline{a_{\mu i s}}\, b_{\lambda\rho i}$$
$$+ \sum_\rho e_\rho \sum_k \left(\frac{\partial b_{\mu\rho s}}{\partial y_k} h_{k\lambda} - \frac{\partial b_{\lambda\rho s}}{\partial y_k} h_{k\mu} \right) = 0 \qquad (\lambda, \mu, s = 1, 2, \cdots, n).$$

These relations will allow us, for example, to express the $e_{\rho i}$ linearly in terms of the e_ρ and the r_1 new arbitrary quantities $\varepsilon_1, \cdots, \varepsilon_{r_1}$, the coefficients being functions of the y's; that is[4]

$$(20) \quad e_{\rho i} = \sum_\sigma^{1,\cdots,r} \alpha_{i\sigma\rho}\, e_\sigma + \sum_n^{1,\cdots,\tau} \beta_{i\lambda\rho}\, \varepsilon_\lambda \qquad (i = 1, 2, \cdots, n;\ \rho = 1, 2, \cdots, \sigma).$$

8. Suppose that for the two expressions of given total differentials, we arrive thus at the same number of invariants $y_1, y_2, \cdots, y_m$ for the first and $Y_1, Y_2, \cdots, Y_m$ for the second; that the coefficients h_{ij} are for the two systems the same functions of their arguments; the same for the coefficients $\overline{a_{iks}}$. Let us determine if the two systems are equivalent, that is to say if we can find for the X's and the U's functions of the x's and the u's which transform the y's into the Y's and the $\overline{\omega}$'s into the $\overline{\Omega}$'s. For this let us assume that the determinant

$$\begin{vmatrix} h_{11} & h_{12} & \cdots & h_{1m} \\ h_{21} & h_{22} & \cdots & h_{2m} \\ \cdots & \cdots & \cdots & \cdots \\ h_{m1} & h_{m2} & \cdots & h_{mm} \end{vmatrix}$$

is not zero, and let us consider the following system of equations:

$$(21) \qquad \begin{cases} Y_1 - y_1 = 0, \\ \cdots\cdots\cdots\cdots, \\ Y_m - y_m = 0, \\ \overline{\Omega_{m+1}} - \overline{\omega_{m+1}} = 0, \\ \cdots\cdots\cdots\cdots, \\ \overline{\Omega_n} - \overline{\omega_n} = 0. \end{cases}$$

If we find for the X's and the U's functions of the x's and the u's satisfying these equations, they will also satisfy, according to the hypothesis made, the equations

$$(22) \qquad \begin{cases} \overline{\Omega_1} - \overline{\omega_1} = 0, \\ \cdots\cdots\cdots, \\ \overline{\Omega_m} - \overline{\omega_m} = 0. \end{cases}$$

[4]In the second summation, n should be λ and in the range for ρ, σ should be r.

Therefore, if we bear in mind the equality of the Y's and the y's, the bilinear covariants of the first terms of the equations to be integrated,

$$\begin{aligned} \overline{\Omega_{m+1}} - \overline{\omega_{m+1}} &= 0, \\ \cdots\cdots\cdots &, \\ \overline{\Omega_n} - \overline{\omega_n} &= 0, \end{aligned}$$

become

$$\begin{gathered} \sum_{i,\rho} b_{i,\rho,m+1}\, \overline{\omega_i}\, (\overline{\Pi_\rho} - \overline{\varpi_\rho}), \\ \cdots\cdots\cdots\cdots, \\ \sum_{i,\rho} b_{i,\rho,n}\, \overline{\omega_i}\, (\overline{\Pi_\rho} - \overline{\varpi_\rho}). \end{gathered}$$

Let us apply the theory of systems in involution.[5] Designate by

$$t_1, t_2, \cdots, t_n;\ t'_1, \cdots, t'_n;\ t_1^{(n-1)}, \cdots, t_n^{(n-1)}$$

n systems of n arbitrary variables and consider the matrix with $n(n-m)$ rows and r columns:

$$\left|\begin{matrix} \sum b_{i,1,m+1}\, t_i & \sum b_{i,2,m+1}\, t_i & \cdots & \sum b_{i,r,m+1}\, t_i \\ \cdots\cdots & \cdots\cdots & \cdots & \cdots\cdots \\ \sum b_{i,1,n} t_i & \sum b_{i,2,n} t_i & \cdots & \sum b_{i,r,n} t_i \\ \sum b_{i,1,m+1}\, t'_i & \sum b_{i,2,m+1}\, t'_i & \cdots & \sum b_{i,r,m+1}\, t'_i \\ \cdots\cdots & \cdots\cdots & \cdots & \cdots\cdots \\ \sum b_{i,1,n} t'_i & \sum b_{i,2,n} t'_i & \cdots & \sum b_{i,r,n} t'_i \\ \sum b_{i,1,m+1}\, t''_i & \sum b_{i,2,m+1}\, t''_i & \cdots & \sum b_{i,r,m+1}\, t''_i \\ \cdots\cdots & \cdots\cdots & \cdots & \cdots\cdots \\ \sum b_{i,1,n} t_i^{(n-1)} & \sum b_{i,2,n} t_i^{(n-1)} & \cdots & \sum b_{i,r,n} t_i^{(n-1)}. \end{matrix}\right|$$

Designate by σ_1 the degree of the principal determinant of the matrix obtained by taking the first $n-m$ rows, by $\sigma_1 + \sigma_2$ the degree of the principal determinant of the matrix obtained by taking the first $2(n-m)$ rows, etc., by $\sigma_1 + \sigma_2 + \cdots + \sigma_n$ the degree of the principal determinant of the matrix obtained by taking all $n(n-m)$ rows.

Finally, we consider the system of linear equations with rn unknowns $z_{\rho i}$

$$(23) \qquad \left\{\begin{aligned} &\sum_\rho b_{\mu\rho,m+1}\, z_{\rho\lambda} - \sum_\rho b_{\lambda\rho,m+1}\, z_{\rho\mu} = 0, \\ &\cdots\cdots\cdots\cdots, \\ &\sum_\rho b_{\mu\rho n}\, z_{\rho\lambda} - \sum_\rho b_{\lambda\rho n}\, z_{\rho\mu} = 0, \end{aligned}\right. \qquad (\lambda, \mu = 1, 2, \cdots, n).$$

[5]See É. Cartan, *Annales de l'École Normale* (3), 21 (1904), pp. 154-175.

The system (21) *will be in involution if the number of unknowns* $z_{\rho i}$ *which may be taken arbitrarily is equal to*

$$\sigma_1 + 2\sigma_2 + 3\sigma_3 + \cdots + n\sigma_n.$$

In this case the system of equations (21) admits an infinite number of solutions depending on σ_n arbitrary functions of n arguments, σ_{n-1} arbitrary functions of $n-1$ arguments, and so forth. If we have a particular solution, the general solution is obtained by operating on the x's and the u's with the most general transformation which leaves the functions y and the expressions $\overline{\omega}$ invariant. This transformation generates a group of which we immediately have the *structure* by the formulas which give $\overline{\omega'_1}, \overline{\omega'_2}, \cdots, \overline{\omega'_n}$.

These hypotheses applied to the coefficients h_{ik} show that we have

$$\sum_i^{1,\cdots,n} h_{ki} b_{\lambda\rho i} = 0 \qquad (k = 1, 2, \cdots, m;\ \lambda = 1, 2, \cdots, n;\ \rho = 1, 2, \cdots, r).$$

Hence we can without difficulty substitute for the $mn(n-1)/2$ linear equations (23) the system of $n^2(n-1)/2$ equations

$$\sum_\rho^{1,\cdots,r} (b_{\mu\rho s} z_{\rho\lambda} - b_{\lambda\rho s} z_{\rho\mu}) = 0 \qquad (\lambda, \mu, s = 1, 2, \cdots, n). \tag{24}$$

We see thus that the number of unknowns $z_{\rho i}$ which we can take arbitrarily is nothing but the number which we have designated by r_1. The condition of involution is thus

$$r_1 = \sigma_1 + 2\sigma_2 + \cdots + n\sigma_n.$$

9. Let us suppose now that the system (21) is not in involution, that is to say that we have

$$r_1 < \sigma_1 + 2\sigma_2 + \cdots + n\sigma_n.$$

We will form the bilinear covariants of the r expressions $\overline{\varpi_\rho}$, which depend on the x's, the u's and the r_1 new auxiliary variables (r_1 of the quantities $\nu_{\rho i}$). In the same manner in which the r_1 quantities entered in the expressions $\overline{\varpi}$, we see that we have formulas of the form

$$\overline{\varpi'_\tau} = \sum_{(ik)} A_{ik\tau} \overline{\omega_i}\,\overline{\omega_k} + \sum_{i,\rho} B_{i\rho\tau} \overline{\omega_i}\,\overline{\varpi_\rho} + \sum_{(\rho\sigma)} C_{\rho\sigma\tau} \overline{\varpi_\rho}\,\overline{\varpi_\sigma} + \sum_{i,\lambda} D_{i\lambda\tau} \overline{\omega_\lambda},$$

the χ's being r_1 new Pfaffian expressions independent of the $\overline{\omega}$'s and the $\overline{\varpi}$'s.

If we apply the *fundamental identity*[6] to the equations (14), we obtain

$$\overline{\varpi'_\tau} = \sum_{(ik)} A_{ik\tau} \overline{\omega_i}\,\overline{\omega_k} + \sum_{i,\rho} \alpha_{i\rho\tau} \overline{\omega_i}\,\overline{\varpi_\rho} + \sum_{(\rho\sigma)} c_{\rho\sigma\tau} \overline{\varpi_\rho}\,\overline{\varpi_\sigma} + \sum_{i,\lambda} \beta_{i\lambda\tau} \overline{\omega_\lambda} \tag{25}$$

$$(\tau = 1, 2, \cdots, r),$$

[6]See É. Cartan, *Annales de l'École Normale* (3), 21 (1904), p. 155.

the coefficients $c_{\rho\sigma\tau}$ being the structure coefficients of Γ, the coefficients $\alpha_{i\rho\tau}$ and $\beta_{i\lambda\tau}$ being those which enter the equations (20).

Thus there remain only the coefficients $A_{ik\tau}$ which may be independent of the y's. If we assume now

$$\overline{\chi_\lambda} = \chi_\lambda + w_{\lambda 1}\overline{\omega_1} + w_{\lambda 2}\overline{\omega_2} + \cdots + w_{\lambda n}\overline{\omega_n},$$

$$\overline{A_{ik\tau}} = A_{ik\tau} + \sum_{\lambda}^{1,\cdots,r_1} (\beta_{i\lambda\tau} w_{\lambda k} - \beta_{k\lambda\tau} w_{\lambda i}),$$

the expressions $\overline{\chi_\lambda}$ and the coefficients $\overline{A_{ik\tau}}$ are the *invariants*. The $\overline{A_{ik\tau}}$ depend now on the n variables x, the r quantities u, the r_1 quantities ν and the nr_1 quantities ω.

As above (no. 7), we show that these coefficients are subject to a group Γ_2 with $r+r_1+nr_1$ parameters. Taking, in effect, the changes δ of the two terms of equations (25), we obtain

$$\sum_{i,\lambda} \beta_{i\lambda\tau}\overline{\omega_i}\,\delta\overline{\chi_\lambda} + \sum \delta\overline{A_{ik\tau}}\,\overline{\omega_i}\,\overline{\omega_k} = \cdots,$$

the second terms being the familiar expressions. We easily deduce from this

$$\delta\overline{\chi_\lambda} = -d\varepsilon_\lambda + \sum_i \varepsilon_{\lambda i}\overline{\omega_i} + \cdots, \tag{26}$$

the unwritten terms of equations (26) being linear in the e's and in the ε's, also in the $\overline{\varpi}$'s and in the $\overline{\chi}$'s, the coefficients not dependent on the y's. As for the $\delta A_{ik\tau}$'s, they are expressible linearly in terms of the A's themselves, the coefficients being the linear expressions in the e's and the ε's with coefficients which are functions of the y's. We have thus the infinitesimal transformations of the group Γ_2 with $r + r_1 + nr_1$ parameters.

10. If the $\overline{A_{ik\tau}}$ depend effectively on certain of these parameters, we can always reduce a certain number of them to constants; the others will then be the invariants (functions only of the variables x) and the group Γ_2 will be reduced to one of its subgroups. The reduction of Γ_2 to one of its subgroups may also reduce either Γ_1 or Γ; the new invariants, if there are any, may also reduce Γ.

In every case we may, after a certain number of reductions and, if necessary, the introduction of new invariants y, assume that we have reduced to the case where the $\overline{A_{ik\tau}}$ are functions only of the invariants y and where the group Γ_2 is reduced to one of its subgroups with $r + r_1 + r_2$ parameters.

We consider now the characteristic numbers $\sigma_1', \sigma_2', \cdots, \sigma_n'$ corresponding to the system of coefficients $\beta_{i\lambda\rho}$. If we have

$$r_2 = \sigma_1' + 2\sigma_2' + \cdots + n\sigma_n',$$

the system of the $\overline{\Omega}$'s will be equivalent to the system of the $\overline{\omega}$'s on the condition that, for this second system, we arrive at the same number m of invariants Y and that the coefficients $h_{ik}, \overline{a_{iks}}, \overline{A_{ik\tau}}$ be the same functions of their arguments Y as for the system of the $\overline{\varpi}$'s. The most general transformation taking the $\overline{\varpi}$'s into the $\overline{\Omega}$'s depends on σ_n' arbitrary functions of n arguments, σ_{n-1}' arbitrary functions of $n-1$ arguments, and so forth. If we have a particular transformation which takes the $\overline{\omega}$'s into the $\overline{\Omega}$'s, the most general is obtained by operating on the x's with a group with m invariants, the *structure* of which is given by formulas (14) and (25).

If we have

$$r_2 < \sigma_1' + 2\sigma_2' + \cdots + n\sigma_n',$$

we will have the bilinear covariants of the expressions $\overline{\chi_\lambda}$; r_2 new expressions θ are introduced, and we may arrange so that all the coefficients of the $\overline{\chi_\lambda}$ are functions of the y's, except the coefficients of the the $\overline{\omega_i}\,\overline{\omega_k}$. We then proceed as before, arriving at a group Γ_3 with $r+r_1+r_2+r_3$ parameters and so forth.

The theory of systems in involution shows that these operations will have an end, that is to say, that for some integer α we will have

$$r_\alpha = \sigma_1^{(\alpha-1)} + 2\sigma_2^{(\alpha-1)} + \cdots + n\sigma_n^{(\alpha-1)},$$

Thus we will have the *structure* of the most general group which leaves invariant the expressions $\overline{\varpi}$ and the degree of indeterminacy of the transformation which takes the system of the $\overline{\varpi}$'s into an equivalent system.

11. More generally, let us consider:

On one hand, m independent functions $y_1, \cdots, y_m$ of the variables $x_1, x_2, \cdots, x_n$ and two systems of n expressions of independent linear total differentials

$$\begin{matrix} \omega_1, & \omega_2, & \cdots, & \omega_n, \\ \theta_1, & \theta_2, & \cdots, & \theta_n. \end{matrix}$$

On the other hand, m independent functions $Y_1, \cdots, Y_m$ of the variables $X_1, X_2, \cdots, X_n$ and two systems of n expressions of independent linear total differentials

$$\begin{matrix} \Omega_1, & \Omega_2, & \cdots, & \Omega_n, \\ \Theta_1, & \Theta_2, & \cdots, & \Theta_n. \end{matrix}$$

We propose to determine whether there exists a change of variables which transforms the functions y into the functions Y, such that the expressions Ω reduce to the ω's by a linear substitution belonging to a given group Γ of order r, and the expressions Θ to the expressions θ by a linear substitution belonging to another given group Γ' of order r', the coefficients of the finite equations of these two groups possibly depending, in addition to the parameters, on the functions y.

If we designate by

$$\begin{matrix} u_1, & u_2, & \cdots, & u_n, \\ v_1, & v_2, & \cdots, & v_n; \end{matrix}$$

the parameters of the groups Γ and Γ', and if we designate by $\overline{\omega_s}$ and $\overline{\theta_s}$ that which ω_s and θ_s become under the most general substitution of the groups Γ

and Γ', it is clear that the $\overline{\theta}$'s may be expressed linearly in terms of the $\overline{\omega}$'s by the formulas

$$\overline{\theta_s} = l_{s1}\,\overline{\omega_1} + l_{s2}\,\overline{\omega_2} + \cdots + l_{sn}\,\overline{\omega_n} \qquad (s = 1, 2, \cdots, n),$$

the coefficients l_{ik} being functions of the x's, the u's, and the v's. If we operate on the $\overline{\omega}$'s with a substitution from the group Γ and on the $\overline{\Theta}$'s with a substitution from the group Γ', the coefficients l_{ik} will undergo the transformations from a linear group with $r + r_1$ parameters. Reasoning as above, we see that we may reduce a certain number of these coefficients to constant values, the others being *invariant* functions only of the variables x. Each of the groups Γ and Γ' is thus reduced to one of its subgroups, and these two subgroups are obviously isomorphic. We can now disregard the expressions $\overline{\Theta}$, and we are led back to the original problem, the group Γ being reduced to one of its subgroups.

12. The preceding general problem presents itself, for example, if the question arises of recognizing the equivalence of two systems of equations of total differentials in $x_1, x_2, \cdots, x_n$ vis-à-vis a finite or infinite given group G. We have from the theory of groups[7] the result that by adjoining if necessary new auxiliary variables, the group is defined by a certain number of invariants $y_1, y_2, \cdots, y_m$ and by n expressions $\omega_1, \omega_2, \cdots, \omega_n$ subject among themselves to the most general linear substitution of a linear group Γ of order r; on the other hand, if

$$\theta_1 = \theta_2 = \cdots = \theta_\nu = 0 \tag{27}$$

the equations are of one of the given differential systems,

$$\Theta_1 = \Theta_2 = \cdots = \Theta_\nu = 0 \tag{28}$$

those of the other (where the variables are written X instead of x), and designating by $\theta_{\nu+1}, \cdots, \theta_n$, $n - \nu$ arbitrary expressions independent of the first ν (the same for $\Theta_{\nu+1}, \cdots, \Theta_n$), we must take the θ's to the Θ's by the most general linear substitution, which transforms the system of equations (27) into the system of equations (28); this substitution generates the group earlier called Γ'.

[7]See É. Cartan, *Annales de l'École Normale* (3), 21 (1904), p. 176.

References

[BC3G. 1989] R. Bryant, S. S. Chern, R. Gardner, H. Goldschmidt, and P. Griffiths, Exterior Differential Systems MSRI series,, Springer-Verlag, Berlin, New York, 1989.

[C. 1908] Élie Cartan, *Les sous-groupes des groupes continus de transformations*, Annales de l'École Normale (1908), pp. 57–194; Oeuvres Complètes de Élie Cartan, Vol. II, Centre National de la Recherche Scientifique, Paris, 1984, pp. 719–856.

[C. 1910] Élie Cartan, *Les systèmes de Pfaff à cinq variables et les équations aux derivées partielles du second ordre*, Annales de l'École Normale (1910), pp. 109–192; Oeuvres Complètes de Élie Cartan, Vol. II, Centre National de la Recherche Scientifique, Paris, 1984, pp. 927–1010.

[C. 1922] Élie Cartan, *Sur un théorème fondamental de M.H. Weyl dans la theorie de l'espace métrique*, C.R. Acad. Sci., 175 (1922), pp. 82–85; Oeuvres Complètes de Élie Cartan, Vol. III, Centre National de la Recherche Scientifique, Paris, pp. 629–632.

[C. 1923] Élie Cartan, *Les espaces à connexion conforme*, Ann. Soc. Pol. Math. 2, (1923), pp. 171–221.

[C. 1930] Élie Cartan, *Sur un problème d'équivalence et la théorie des espaces métriques généralisés*, Mathematica, 4, (1930), pp. 114–136.

[C. 1934] Élie Cartan, *Les problèmes d'équivalence*, Oeuvres Complètes de Élie Cartan, Vol. III, Centre National de la Recherche Scientifique, Paris, 1984, pp. 1311–1334.

[C. 1951] Élie Cartan, La théorie des groupes finis et continus et la géométrie différentielle, Gauthier-Villars, Paris, 1951.

[Ch. 1940] S. S. Chern, *The geometry of a differential equation* $Y''' = F(X, Y, Y', Y'')$, Sci. Rep., Tsing Hua Univ, 1940, pp. 97–111.

[Ch. 1954] S. S. Chern, *Pseudo-groupes continus infinis*, Colloque de Géom. Diff., Strasbourg, 1954, pp. 119–136.

[Ch. 1955] S.S. Chern, *An elementary proof of the existence of isothermal parameters on a surface*, Proc. Amer. Math. Soc. 6 (1955), pp. 771–782.

[Ch. 1957] S. S. Chern, *On a generalization of Kähler geometry*, Algebraic Geometry and Topology, Princeton, NJ, 1957, pp. 103–121.

[Ch. 1966] S. S. Chern, *Geometry of G-structures*, Bull. Amer. Math. Soc., 72 (1966), pp. 167–219.

[Chr. 1869] E. B. Christoffel, *Über die Transformation homogener Differentialausdrucke zweiten Grades betreffendes Theorem*, J. Reine Angew. Math. (Crelle), 70 (1869), pp. 46–70, pp. 241–245.

[F. 1963] H. Flanders, Differential Forms with Applications to the Physical Sciences, Academic Press, New York, 1963.

[G. 1983] Robert Gardner, *Differential geometric methods interfacing control theory*, Progress in Mathematics, Vol. 27, Birkhäuser, Boston, 1983, pp. 117–180.

[G. 1985] Robert Gardner and William Shadwick, *Equivalence of 1-dimensional La-*

grangian field theories in the plane I, Lecture Notes in Mathematics 1156, Springer-Verlag, Berlin, New York, pp. 154–178.

[G-S. 1986] ROBERT GARDNER AND WILLIAM SHADWICK, *Overdetermined equivalence problems*, Differential Geometry: The Interface Between Pure and Applied Mathematics; Proceedings of a Conference held April 23–25, 1986, American Mathematical Society Contemporary Mathematics, 68, Providence, RI, 1986, pp. 110–120.

[G-S. 1987] ROBERT GARDNER AND WILLIAM SHADWICK, *Feedback equivalence of control systems*, Systems Control Lett., 8 (1987), pp. 463–465.

[Gr. 1983] P. GRIFFITHS, Exterior Differential Systems and the Calculus of Variations, Progress in Mathematics, 25, Birkhäuser, Boston, 1983.

[Gr-J. 1987] P. GRIFFITHS AND G. JENSEN, *Differential Systems and Isometric Embeddings*, Ann. of Math. Stud. 114, Princeton University Press, Princeton, NJ, 1987.

[K. 1921] F.KLEIN, *Gesammelte Mathematische Abhandlung*, Vol. 1, J. Springer, Berlin, 1921, pp. 460–.

[K-N. 1965] S. KOBAYASHI AND T. NAGANO, *On a fundamental theorem of Weyl-Cartan on G-structures*, J. Math. Soc. Japan, 17 (1965), pp. 84–101.

[P-R-S. 1979] F. PIRANI, D. ROBINSON, AND W. SHADWICK, *Local Jet Bundle Formulation of Bäcklund Transformations*, Math. Phys. Stud., Vol. 1, D. Reidel, Dordrecht, the Netherlands, 1979.

[R. 1953] B. RIEMANN, *Über die hypothesen Welche der Geometrie zu Grunde Liegen*, Collected Works of Bernard Riemann, H. Weben, ed., Dover, NY, 1953, pp. 272–287.

[Si-S. 1965] I. SINGER AND S. STERNBERG, *The infinite groups of Lie and Cartan*, J. Analyse Math., 15 (1965), pp. 1–114.

[S. 1964] S. STERNBERG, *Lectures in Differential Geometry*, Prentice Hall, Englewood Cliffs, NJ, 1964.

[V. 1927] O.VEBLEN, *Invariants of Quadratic Differential Forms*, University Press London, London, 1927.

[Wa. 1971] F. WARNER, *Foundations of Differentiable Manifolds and Lie Groups*, Scott, Foresman, Glenview, IL, 1971.

[W. 1922] H. WEYL, *Die Einzigartigkeit der Pythagoreischen Massbestimmung*, Math. Z., 12 (1922), pp. 114–146.

[W. 1938] H. WEYL, *Cartan on groups and differential geometry*, Bull. Amer. Math. Soc. (1938), pp. 598–601.

[Wi. 1987] G. WILKENS, *Local feedback equivalence of control systems with* 3 *states and* 2 *control variables*. PhD dissertation, University of North Carolina, Chapel Hill, NC, 1987.

Index

Abstract computation, 24
Adjoint representation, 36
Automorphism of the e-structure, 61, 105

Betti number, 73
Bilinear covariants, 109, 111, 113, 118, 119, 121

Cartan, Elie
  equivalence problem of, 1
  general equivalence theorem of, 74
Cartan form, 52
Cartan formula, 43
Cartan–Kähler theorem, 73, 75, 93
Cartan's lemma, 20, 78, 86, 90, 101
Characteristic variety, 75
Christoffel form problem, 69
Christoffel symbols, 68
Coadjoint action, 42, 43
Coadjoint representation, 36, 105
Complete e-structure, 72
Completely integrable system, 16
Conformal geometry, 2, 12, 76, 83, 97, 102
Constant torsion, 65
Contact geometry of a third-order ordinary differential equation, 5
Contact transformation, 7, 8, 76
Contragredient representation, 35, 41

dSS^{-1}, 14, 15, 23, 26, 28, 68
Determinantal action, 42
Diagonal subgroups Δ, 16, 17
dim $\mathcal{G}^{(1)}$, 74, 76, 78, 90
Discrete arithmetic subgroups, 73

e-structure, 57–59, 61, 65, 69, 71, 72, 93, 101, 104–107
  automorphism of, 61
  complete, 72
  equivalence of, 59, 66, 68
  lifted, 11, 19
  order of 58, 59, 69
  rank of 58, 59, 69
Equivalence problem, 53, 77, 83, 84, 88
  of e-structures, 59
  of Elie Cartan, 1
  overdetermined, 53, 54
exp tX, 43

Feedback diffeomorphisms, 81
First fundamental theorem of Lie, 14
First Lie algebra prolongation, 31
First structure equations, 34
First Whitehead theorem, 73
First-order constant type, 37, 38
First-order Lagrangian, 6
First-order normalization, 39, 102
Foliation, 65
Frobenius theorem, 16, 60, 67, 73
Fundamental identity, 119
Fundamental theorem of pseudo-Riemannian geometry, 27

G-space, 11, 22
$G(m)$, 83
G_2, 107
$\Gamma(m,p)$, 83
General Equivalence Theorem of Cartan, 74
Generalized geometry, 89, 94
Generalized orthogonal groups $SO(p,q)$, 26
Geometry of control systems under feedback, 8, 81, 83
Geometry of the integral, 6, 50
Geometry of time-fixed Newton's equations, 4
Group-fiber representation, 19, 20
Groups up to right multiplications, 59

H-reduction of type S, 46
$h^0(\mathcal{G})$, 32
$h^{0,2}(\mathcal{G})$, 32
Hemihedrally isomorphic, 116
Hilbert Invariant Integral, 52

Identification of group of automorphisms on an e-structure, 105
In involution, 74, 76, 90, 92, 94, 97, 119
Independence conditions, 75
Infinitesimal generators, 19, 67
Infinitesimal group actions, 40, 85
Integrable e-structure, 71, 72
Interior product, 43
Intrinsic torsion, 32
Invariant functions, 24, 53, 116
Involutive, 74

Levi–Civita connection, 68
Levi–Civita connection form, 29
Lie algebra valued compatible absorption, 21, 26, 27, 31, 33, 34

Lie derivative, 43
Lie theorems
first fundamental, 14, 66
second, 19
third fundamental, 75
third, local converse to, 17
Lie's theorem only $y" = 0"$, 78
Lifted equivalence problem, 11, 19
Local converse to the Third Lie Theorem, 17
Loop A, 47

$M(n,\mathbf{R})$-valued right invariant Maurer–Cartan form, 15
Maurer–Cartan equations, 16
Maurer–Cartan forms, 14–17, 19–21, 24, 26, 28, 43, 73
set of, 14
T_e-valued, 14
Maximal Abelian subalgebras, 105
Mod base, 19, 20, 22
Modified forms, 23

Natural absorptions, 51, 88
Natural "covariant derivatives," 57
Noncompact real form, 107

One-dimensional time-fixed Newton's equations, 47
One-parameter group of left translations, 43
Order of the e-structure, 58
Overdetermined algorithm, 8, 54, 81
Overdetermined equivalence problems, 53, 54
Overdetermined problem, 3–5, 81
Operator L, 22

Parametric calculation, 25, 52, 89, 102, 103
Particle Lagrangians, 7, 52, 83
$\Pi_{\mathcal{G}}$, 31, 32, 39
Pfaffian equation, 13, 18, 119
Poincaré lemma, 24
Principal components of order one, 20, 21, 49, 51, 97
Principal left G-bundles, 11
Pseudo-Riemannian metric, 94

Quasilinear systems in involution, 73

Rank of the e-structure, 58
Reduced Cartan characters, 74, 76, 77, 90, 98
Reduced tableau matrix, 74, 90, 91
Reduction of the structure group, 38, 53, 54, 65
Regular of rank ρ, 58
Répère Mobile, 39
Restricted equivalence problem, 57
Riemann–Christoffel curvature matrix, 29
Riemann–Christoffel curvature tensors, 69
Riemannian geometry, 1, 28, 57, 68
Right invariant vector field, 43
Roots, 106, 107

S_3-Cartan lemma, 27, 28
S_3-lemma, 27, 69, 98, 100
$SO(p,q)$ absorption theorem, 27, 29, 97
Second Lie theorem, 19
Semisimple Lie algebras, 105, 106
Set of Maurer–Cartan forms, 13
Similarity representation, 41
Stabilizer type, 43
Standard jet coordinates, 4, 6
State estimation of plants under feedback, 81
Structure constants of G, 14
Symbolic manipulation program, 25, 50

τ_0-modified coframe, 38
T_e-valued Maurer–Cartan form, 14
Tableau matrix, 74
Technique of the graph, 18, 59, 62, 64, 71, 72
Tensor product of two representations, 41
Third fundamental theorem of Lie, 75
Third-order differential equation
contact geometry of, 5, 13
Third-order ordinary differential equations under contact transformations, 5, 102
Time-critical closed loop control, 87
Time-fixed Newton's equations, 4, 55
one-dimensional, 47
Torsion coefficients, 21, 107

Torsion terms, 21, 23, 77, 91, 92, 104, 105
Total differential equation, 13
Transitive pseudogroup, 75, 76, 78
Translational action, 42

Underdetermined problem, 13, 81

Web geometry, 2, 22, 54, 57, 65

Zeroth-order invariant, 53